THE PHYSICS OF SPINOZA'S GOD

The Universe Explained

Jeffrey Marshall Webster

1

Why is there something rather than nothing, and why are we here to ask that question? These two queries sum up the riddle of human existence. God, evolution, the Big Bang, aliens – possible sources, but not really explanations. Where did God come from; how did life emerge in the first place before it evolved; what caused the Big Bang; aliens, like us, had to have had an origin at some point. We speak in undefined terms: space, time, matter and energy. The genius of Einstein was altering our preconceptions about these four basic terms. Space is not a void, but has an "independent structure"; time is not absolute, but relative; matter and energy are really the same thing. Einstein made us conceive of these terms differently, but they still remain undefined, inscrutable.

Existence is a riddle. But there are clues to the answer to this riddle. The first clue is that we humans have the faculty to discern the riddle, and, hopefully, solve it. The human brain weighs about three pounds, has eighty billion neurons and 100 trillion synapses. It is the most complex entity we know of. Those who take a strict Darwinian view of the universe would have us believe that this incredible structure is the product of random chemical reactions and natural selection. Those of a religious bent would postulate a transcendent, supernatural Creator as the source not only of humanity, but of the universe itself. In physics there is an idea known as the anthropic principle. In comes in two forms, the weak and strong. The weak version asserts that the fact that we humans exist demonstrates that the universe is conducive to the appearance of life. Common sense, somewhat tautological.

The strong version makes the bold claim that the constants of physics were deliberately fine-tuned to give rise to intelligent life, that intelligence is the very *raison d'etre* of the universe. But that is just a restatement of the religious explanation couched in scientific phraseology.

It is not just the possibility that life and the universe were engineered by some cosmic intelligence that beguiles and enchants us, that gives rise to the great religions and the metaphysics of the great philosophers, but we are vexed by the simple enigma of existence itself. Something in our psyche demands to know how something could have arisen out of nothing, but something else in our psyche is certain that something cannot arise from nothing. Is there even such a thing as nothingness?

At the limits of our understanding is the concept of God, a being that exists outside the universe of time and space, cause and effect. God begins where our reasoning faculty breaks down. We have all had that conversation: "Where did everything come from?"

"God made it."

"Where did God come from?"

"He just always was."

Aristotle conceived of God as the Unmoved Mover, a first cause that somehow brought the material world, with the Earth as its center, into being. Strongly influenced by Aristotle, the Christian theologians saw God as a perfect, all-powerful being whose essence was incomprehensible to the human mind. Like Aristotle, they also conceived of God as transcendent, as outside of and above nature. For over a millennium philosophy would have to content herself with being the handmaid of theology. But when a Polish monk named Copernicus demonstrated that the earth is not the center of the universe, the handmaid found herself suddenly liberated, but not without some difficulties.

In 1600, a former Catholic priest named Giardano Bruno was burnt at the stake in Rome for advocating the Copernican heresy, even going so far as to assert that there were many worlds in the cosmos and that they were inhabited by humans. Thus was born that century of centuries, the 1600's, the age of Galileo and Newton, Descartes and Hobbes – and Spinoza.

One could say that Spinoza was born to heresy. The descendant of Sephardic Jews who had fled Portugal during the Inquisition, Baruch Spinoza was born in Amsterdam, Holland, a tolerant, prosperous city where Jews were free to practice their faith, and the new ideas engendered by the Scientific Revolution and the Enlightenment were in the air. Highly intelligent, Spinoza studied mathematics and philosophy, being particularly influenced by the ideas of Descartes, Giardano Bruno, and the medieval Jewish philosophers Maimonides, Hasdai Crescas, and Levi ben Gerson. Boldly, Spinoza rejected the idea that the Torah was revealed divine truth. He proclaimed the pantheistic view that God and the universe are one. In his view, the terms God and nature are interchangeable. God is the primary substance of the universe, and all the myriad entities we perceive in nature are just some of the infinite modes of that primary substance. As a result of his teachings, Spinoza was excommunicated from the Jewish faith.

Three hundred years later Albert Einstein, when asked if he believed in God, answered that he believed in Spinoza's God, which revealed itself in the orderliness of natural law. Einstein defined his religion as the awe he felt when contemplating the mystery that lay behind our very limited understanding of the universe. However, in a private letter, he remarked that he never felt the need for "the hypothesis of God" to explain the laws of physics and the origin of the universe.

Ironically, Georges Lemaitre, a physicist who also happened to be a Jesuit priest, using Einstein's general relativity as a basis, formulated what would come to be known as the Big Bang theory. Lemaitre, perhaps because of his religious background, assumed that the universe had to have had a specific origin in time. The Big Bang theory came to be, and remains, the generally accepted theory of the scientific orthodoxy. Some maverick theorists who rejected the Big Bang theory found themselves ridiculed, and in one case shunned by the scientific establishment. Science, as the Catholic Church had done three centuries earlier, found it necessary to punish heretics.

The Steady State theory, and a variation thereof known as the Continuous Creation theory, assert that the universe has always existed. According to the latter theory, matter is created from energy in the dense cores of certain types of galaxies and spewed forth, creating new galaxies. But how can matter be created? What is it created from?

The great physicist Richard Feynman once asserted that there is enough energy in a volume of space the size of a teacup to boil all the oceans of the earth. This energy is called zero point energy because when space is chilled to absolute zero degrees, there is still discernible energy within it. Stated another way, space is not a void, but a dense plenum of energy. Light can be understood as a transverse wave in this medium, and electrons and protons as spherical standing waves.

But what is the nature of this energy, this zero point energy that comprises space? As we shall delve into below, there is good reason to believe that zero point energy has a wavelength of 10^{-35} meters, a quantity known in physics as the Planck length. This is an extremely short distance – *twenty powers of ten shorter than the diameter of a proton.* Further, there is reason to assume that zero point energy travels at *twenty billion times the speed of light.* Space is an infinite pool of dense, dense energy. Is this energy the primary substance hypothesized by Spinoza, what he called God?

There is an interdisciplinary theory called "emergence." It is related to the theories of "dynamical systems," "spontaneous order," and "synergetics." Simply put, these theories explain how order can arise out of seeming chaos in systems where there is a steady input of energy. Consider a system like the zero point field, infinite in energy, infinite in time, infinite is space. A configuration of the utmost complexity could arise in such a situation – and what is more complex than intelligence? Imagine: an infinite, unfettered intelligence.

Maybe Spinoza was on to something. And maybe mankind was engineered by this cosmic intelligence. Consider: once we humans learn how to tap into zero point energy, we will have unlimited energy and the ability to communicate, compute, and travel at twenty billion times the speed of light.

Let us consider the possibilities.

2

Consider the human condition. We are born into a universe whose origin is a mystery, whose nature is still debated by physicists and astronomers, which at times seems downright hostile to human existence, and leaves us perplexed as to why – if there is a "why" – we are even here. "What is the meaning of life?" many have asked. Tomes have been written in an effort to answer that question. The entire philosophy of existentialism proceeds from that very question. One can read through the ponderous works of Kierkegaard and Sartre, and come away just as perplexed as before.

Socrates taught that "man is a microcosm of all the mysteries of the universe." One thing is for sure: man is a being whose lot it is to ponder mysteries, the greatest of which is his own existence. Aristotle taught that form follows function. Humans have incredibly complex brains, walk upright, and have opposable thumbs. Therefore, our function – our purpose – seems to be to create. But to create what? To answer that question, one need only to look at what man has already created. We have split the atom, traveled to the moon, and sent unmanned space probes to interstellar space. This is a strong clue to our purpose.

In his treatise *Metaphysics,* Aristotle explained that any entity can be understood in terms of its "causes." Aristotle's concept of cause was slightly different from our understanding of the term. He taught that everything has four causes: material, formal, final, and efficient. For example, the material cause of the Parthenon was the marble with which it was built; the formal cause was its architecture; the final cause was the purpose for which it was built, as a temple to Athena; and the efficient cause was the effort of the architect, laborers and craftsmen who constructed it. In trying to define man in the context of the four causes, the material and formal are rather evident to our senses. About the final and efficient we are left to speculate.

One of the key defining conditions of our nature is the scale on which we exist. We are too large to observe directly the atoms of which we are composed, and too small to comprehend fully the machinations of the galaxies and the clusters thereof. As great as our reasoning power is, as prodigious as the efforts of our greatest thinkers have been, we are still limited by our perspective and our modes of perception. Even Aristotle, as seminal a philosopher as he was, thought the Earth was flat and lay at the center of the universe. Because he could not explain the origin of mankind, he postulated that man had always existed. In his *Physics,* he simply accepted as a given the commonsense notion that heavier objects fall faster than lighter ones. This idea was accepted without question for nineteen centuries until it was disproved by Galileo, who dropped a one pound weight and a ten pound weight from the Leaning Tower of Pisa and, when the weights hit the ground at the same time, proved that, regardless of weight, objects fall at the same speed. In the same century that Galileo demonstrated that objects fall at the same speed regardless of weight, Isaac Newton asked why things fall at all. He demonstrated that objects exert a force on other objects that decreases as the square of the distance between them. He also showed that this force – gravity – causes falling objects to accelerate at 32 feet per second. To this day, physicists are still debating what causes gravity.

Newton also discovered that light, when passed through a prism, is split into the different colors of the rainbow. He reasoned that light is composed of little particles he called "corpuscles." However, in the next century, Thomas Young, in a famous experiment, demonstrated that light travels as a wave. But this raised a question. A wave, by definition, is a periodic disturbance in a medium that carries energy from one point to another without permanently altering the medium. So what was the nature of the medium through which light traveled? Some thinkers postulated the existence of a "luminiferous ether" which pervaded space. However, other thinkers theorized that space was not so much a void pervaded by a medium, but was a medium itself, that it had its own structure. Lord Kelvin postulated that space is made up of interlocking tetradecahedrons – fourteen-face geometric solids. In the nineteenth century, William Kingdon Clifford theorized that matter itself is wave-like phenomenon in the medium of space.

The miracle of the process which began in the seventeenth century is that humans used reason, empirical observation, and experiment to see beyond the perspective in which we were trapped, and the modes of perception by which we were limited. The earth was no longer flat and at the center of the universe; the nature of gravity had been captured in precise mathematics; and space was no longer a void and an abstraction, but a thing in itself, although beguiling in nature.

Understanding space, finding a way to perceive space, is the key to solving the riddle of human existence, indeed, of the universe itself. Einstein, rejecting the idea of the ether as an unnecessary postulate, asserted that space itself had a structure that could be curved by matter. It was this curvature that explained the phenomenon of gravity. In Einstein's view, gravity was not so much a force as the geometry of space-time. Matter in motion follows the curvature of space. But what is matter? Indeed, what is motion? As we shall see, Einstein helped us look beyond our perspective to understand these two phenomena as well.

So, what is man? Among other things he is a creature who can look beyond the sometimes illusory nature of commonly held notions and see the true reality beyond. If it is true that he who overcomes himself is divine, then perhaps there is a spark of the divine in mankind. But what is this entity, or state of being, that we call divine? If we were indeed designed and engineered by some being or beings greater than ourselves, then what was their motive? Further, on how many other planets has this same process occurred? Were we engineered for the purpose of utilizing the resources on our planet to contrive a way to travel to other planets, other galaxies, and start the process anew? Are we the *andressin* of some superior being or beings?

Andessin is a German word that be translated as "otherness." Are we the pawns of superior beings, or perhaps their vessels? Is mankind the ultimate *ding an sich* of the universe, a phenomenon that exists for its own sake, needing no further meaning or justification? Are we an accident of chemistry and circumstance? Even if it were true, it would not negate the fact that we look toward the far-flung galaxies with longing.

One of the most, if not the most, demonstrably intelligent men who ever lived was the great Hungarian-born polymath John von Neumann. He helped develop both the atomic and hydrogen bombs, invented the first stored-program computer, co-invented game theory, and made many other advances in diverse fields. In pondering the future of mankind, he speculated that in the future we would develop space probes that would have the ability of traveling to other planets and using the raw materials there to make replicas of themselves that would travel to other planets and repeat the process *ad infinitum,* thus spreading intelligence throughout the universe. These hypothetical craft are called, fittingly, von Neumann probes. But suppose that we humans are not just the ones who will build those probes, but *are* those probes. Perhaps John von Neumann was the ultimate von Neumann probe. During his lifetime, when he associated with such luminaries as Einstein, Oppenheimer, and Godel at the Institute for Advanced Studies in Princeton, there were people who, only half-jokingly, speculated that someone as smart as von Neumann could not be human, that he was really an extraterrestrial. Who knows?

But even von Neumann had his limitations. When shown the schematic of what would eventually become a laser, he said that the idea would never work. As we shall see, sometimes even the best and brightest among us are limited by ingrained ideas, by the limited perspective that hobbles all us humans. One such idea is the idea of the supernatural, the assumption that there exists a level of being outside and above nature that is beyond the ken of human understanding. Alexander Pope, a contemporary and friend of Isaac Newton, expounded on this view in his long poem "Essay on Man." A devout Catholic, he felt that humans should avoid speculations about God and accept on faith revealed scripture. Mankind, in his view, should tend to earthly matters. He summed up this attitude in the maxim: "The proper study of Mankind is Man."

Even Spinoza thought there were attributes of God humans could not grasp, but, unlike Pope, he did not believe in the duality of the natural world and the supernatural. In his view the term supernatural is a contradiction in terms. The German philosopher Hegel seemed to share Spinoza's view. Indeed, he seems to have been greatly influenced by him. Hegel conceived of God as a being that unfolds in nature. He called this being the *Weltgeist,* which can be translated as either World Spirit or World Idea. In his postulate "what is rational is real and what is real is rational," Hegel asserted that human reason can comprehend the *Weltgeist* by studying its manifestations in nature and human history, its "phenomenology." Not surprisingly, Hegel had a strong influence on certain theologians. Philosophy was no longer just a handmaid to theology after Hegel.

Of course, the Scientific Revolution and the Enlightenment also engendered atheistic thinkers like Thomas Hobbes and Baron d'Holbach who believed that the universe, including human intelligence, could be explained in purely materialist terms, as arising from the random machinations of matter and energy. This view was reinforced in 1859 when Charles Darwin published his theory of evolution. After that, atheistic materialism became the vogue among intellectuals, so much so that Nietzsche famously opined "God is dead." The irony is that atheists can be as emotionally dogmatic about their disbelief as the most ardent believers. And therein lies the rub. We must set aside the emotion as well as the preconceptions and look at the universe anew. We must see what is there, not what we want to be there – or not there.

Like good detectives, let us first collect the evidence before we develop a theory of the case.

3

And God said," Let there be light," and there was light. – Genesis 1:3

*"I would like to spend the rest of my life thinking
about what light is."* – Albert Einstein

Light. Just the very word evokes images of the sun, the moon, and the stars, of the abstractions truth, beauty, and salvation. Ask a painter or a poet to express light, and you may get a picture of a sunrise over the mountains, or a poem about a beacon on a rocky coast guiding a ship to shore. In some religions lighting a candle is an act of supplication to the Divine. Before proclaiming Christianity as the official religion of the Roman Empire, Constantine and his soldiers worshipped *Sol Invictus* – the Unconquerable Sun. Jesus said, "I am the light of the world." Before Satan was cast out of heaven for disobedience, he was the beautiful angel Lucifer – Light Giver.

We think of light in terms of rays, shafts, and beams, as an ethereal flow of warm beneficence washing over our world. With the advance of biological science, we learned that
green plants actually use sunlight to make nourishment. It was if science had vindicated the worshippers of *Sol Invictus* after all. Yet light for most of us is an undefined term, an abstraction. At best it is a mystery. So what exactly is light?

When we speak of light most people think of visible light. But there are other forms of light – gamma rays, x-rays, ultraviolet radiation, infrared light, microwaves, radio waves, and radar. Visible light is in the middle of the spectrum between ultraviolet and infrared. What distinguishes one form of light from another is the wavelength. Gamma rays have a very short wavelength, and hence carry much more energy than visible light. On the other end of the spectrum, radar has a very long wavelength and is much less energetic that visible light.

Light is electromagnetic radiation. By definition, electromagnetic radiation is alternating electric and magnetic fields that travel through space at right angles to one another. A wave that moves in this manner is called a transverse wave. Interestingly, transverse waves are associated with solid mediums. Solids are made up of atoms arrayed in a lattice-like manner. Thus, the right-angle polarization of the transverse wave structure is explainable by the very structure of the solid. But how does this explain the transverse wave structure of light? This was one of the things Einstein wanted to spend the rest of his life thinking about.

To further compound the mystery of light, the great physicist Max Planck, in a classic experiment in 1899, discovered that light is not emitted in a continuous flow – a beam – but in discrete units or "quanta." This begs the question: if light is a wave, how can it also be a discrete unit, a particle? Someone once offered the analogy of individual soldiers marching in a line. Einstein was perhaps more accurate in comparing light to wave on the ocean, and light quanta to the atoms that make up the water.

In 1905 Einstein published five papers that revolutionized physics. Two of them introduced the concept of relativity. Another dealt with the photoelectric effect, the phenomenon in which light stimulates the flow of electrons in certain materials. In this paper Einstein demonstrated mathematically that light had to have a particulate nature in order to induce such a flow. He called these light units "photons."

Light is also a messenger. Our eyes, by definition, are organs that detect electromagnetic radiation of wavelengths 400 – 800 nanometers. Because we perceive objects, and the movement thereof, by means of the light emitted by them, certain strange anomalies occur when those objects approach the speed of light. Einstein considered these anomalies in his first paper on relativity. The strangest of these anomalies, and the hardest to understand, is time dilation. Einstein showed mathematically, and it has been proven experimentally, that time moves slower in a frame of reference that is moving at a greater velocity than the one against which it is measured. If a man left Earth on a spacecraft traveling 99.99% of the speed of light and returned two years later by his clock, he would discover that 200 years had passed on earth.

The reason so many people have difficulty understanding time dilation is that physicists have failed to provide an adequate definition of time. Even Einstein treated time as an undefined term. Its nature was just assumed to be self-evident. Some people think of time not only as a measurement, but as a dimension in and of itself. Indeed, there is more than one way to define time. For example, we could define a chunk of time as those set of events that happened between event A and event Z. *Events.* This is the key word. Time is a *quantity* of *change.* Very well, but what is change? The answer to this question is implied by one of the central corollaries of relativity set down by Einstein: there is nothing in the universe in a state of absolute rest. This is why motion has to be measured relative to another object which, of necessity, is also in motion. Stillness is only an illusion. You may think you are sitting still in your easy chair, but you are on a planet that is spinning at over a thousand miles an hour, is orbiting a star that is orbiting the center of a galaxy that is moving through space toward some mysterious gravitational attractor. *There is no such thing as stillness.* And this is the key to understanding everything: *to move is to be, to be is to move.* Stillness is an illusion, an artifice of perception and cognition – just as nothingness is.

Let us consider the example of the space traveler who, by his clock, was only gone from earth two years. How is it that two hundred years had passed on earth? Because time is a measurement of change, and change is motion, as much change occurred in the space traveler's frame of reference as occurred on earth because the space traveler's frame of reference underwent a greater quantity of motion than the earthbound frame of reference. Another way to visualize this is by representing the earthbound frame of reference as a reel of film. In between each frame is a little gap that the shutter of the projector blacks out. Now represent the space traveler's frame of reference with a reel of film the same size as the other, but now picture the gaps between the frames as being twenty-five feet long. If you ran the two reels of film on two projectors side by side that were revolving at the same speed, the second reel would only project a small fraction of the events projected by the first, but the same quantity of film would roll on each projector. And, as this analogy aptly implies, *neither time nor space are continuums, but are composed of discrete units.* We will discuss this later in greater detail.

In his second paper on relativity, Einstein also considered the nature of matter and energy. He pondered the question whether the mass of a body is a measure of its energy content. He wondered if a piece of radioactive material would lose mass as it radiated energy. His ruminations led him to a conclusion that

he would later express in his famous equation $E = mc^2$: matter and energy are really the same thing. The duality between the two is another illusion wrought by our perspective and modes of perception. But how can this be?

Consider the phenomenon known as pair creation. When a high-energy photon comes near a heavy atomic nuclei, or passes through a strong magnetic field, it spontaneously transforms into an electron and a positron, the latter being the same mass and dimensions of an electron, but the opposite electric charge. But how can a wave become a particle? It depends on how you define particle. The physicist Milo Wolff made a very compelling case that electrons, and all other subatomic particles, are spherical standing waves in the medium of space. A standing wave is by definition two waves moving in opposite directions superimposed on one another. For example, a guitar string, when plucked, creates a two dimensional standing wave that imparts vibration to the surrounding air molecules. A spherical standing wave is, of course, in three dimensions.

Light radiates outward. But in those special circumstances where space is highly curved for some reason, it becomes trapped and goes around and around, back and forth. Consider that light moves at 186,000 miles per second. That means that a photon could orbit the earth over seven times in one second. Imagine that same photon chasing its tail in a space one trillionth of a meter wide and you have a good idea of how energy could, to beings on the human scale, be construed as a solid object.

Thus, all we have to do is make a few adjustments in the way we visualize things, the way we conceive things, and we come to understand that physical reality is not only comprehensible, but, at base, rather simple.

Einstein's ultimate dream was to derive a "unified field theory," a set of a few basic equations that would explain the force that lay at the base of all other forces of physics. At this he failed. As we shall see, physical reality often gives rise to phenomena that are totally unpredictable by the initial conditions from which they arise. The whole is often more – and quite different – from the sum of its parts.

4

Nowadays we accept the existence of atoms as a given. But just a little over a century ago there were still those who rejected the idea. In 1905, in yet another one of his papers from his miracle year, Einstein demonstrated the existence of atoms in his investigation of Brownian motion, the phenomenon whereby grains of pollen suspended in water randomly jiggle. Einstein explained this motion as being caused by the collision of water molecules with the pollen grains. However, it would not be until 1913 that Niels Bohr put forth the model of the atom that still holds toady: a nucleus composed of protons and neutrons orbited by electrons.

The existence of atoms had been theorized since ancient times. In the fifth century BC, the Greek philosophers Democritus and Leucippus asserted that matter is composed of four elements – air, earth, water, and fire – and that each element is composed of indivisible particles. The word "atom" derives from the Greek word for "indivisible." Plato taught that the atoms of each element had a distinctive geometric shape. Earth atoms were cubes, fire atoms tetrahedrons (four-faced polyhedrons), air octahedrons (eight-faced), and water icosahedrons (twenty faced). Plato's logic is that all these geometric solids were perfect in the mathematical sense because
each of their faces were identical and all their edges and angles were congruent.

The Greek philosophers were well versed in mathematics, and much of their speculations about nature were based on the idea that nature would mirror the orderliness, logic, and exactitude thereof. But not all of them accepted the atomic hypothesis. Aristotle believed that the four elements were infinitely divisible. In addition, he postulated the existence of a fifth element, a "quintessence," of which the heavenly bodies were composed. He conceived of this fifth element as superior to the four mundane elements.

Consider the logic behind the atomic theory. Simply put, matter is not infinitely divisible. Infinity does not extend inward to the small, but outward toward the cosmos. Implicit in this idea is that space itself is not infinitely divisible. Another concept that logically emerges from the hypothesis is that atoms are by definition particles incapable of further division, that they are irreducible solids. This idea of solidity was the basis of a notion accepted as a given for over two millennia: the duality of matter and energy, that they were two intrinsically different phenomena. As we have seen, Einstein destroyed this dichotomy.

So what are atoms, and how did they come to be? We know that electrons can be created from photons and vice versa. But what about protons, the positively charged constituent of the atomic nucleus, and neutrons, equal in mass to a proton, but electrically neutral?

The first riddle of the proton is that, although it is 1,834 times more massive than an electron, it has the same amount of electric charge, albeit opposite charge, as the lighter particle. A curious property of the neutron is that, outside the atomic nucleus, it will, in about eleven minutes, decay into a proton and an electron. Thus, a neutron can be defined as a composite entity composed of a proton and an electron. So, that leaves us with the question: how are protons created? We know that stars are powered by the fusion of hydrogen nuclei to form helium nuclei, and that heavier elements are formed by nuclear fusion in the cores of stars. But how were the protons that make up hydrogen nuclei formed? If one accepts the existence of the zero point field, the explanation is rather simple. But before we turn to that question, let us delve into the Big Bang theory of the origin of matter, and its heterodox rival, the Steady State theory.

In the early years of the twentieth century when Einstein was formulating his theories of gravity, the Milky Way was considered the entire universe. Telescopes were not yet strong enough to detect the billions and billions of galaxies that make up the known universe. Einstein's calculations suggested two possibilities: the universe was either expanding or contracting. Rejecting either possibility, Einstein derived a mathematical device known as the "cosmological constant" which made his equations adhere to a model of a static universe. However, Georges Lemaitre seized upon the possibilities in Einstein's theories to formulate his theory that the universe originated in a single point, a "primitive atom," from which all matter and energy, and space and time emerged. The term Big Bang was coined in the 1940's as a derisive term by the astronomer and physicist Fred Hoyle, who believed that the universe had always existed.

The Big Bang theory came to have wide acceptance in the physics community in large measure due to the observations of astronomer Edwin Hubble, after whom the Hubble telescope is named. In 1924, his observations led to the discovery that the Milky Way did not comprise the entire universe, that there were other galaxies. Further, he observed that the spectra of the light from some galaxies were red-shifted. This means that the pattern of lines, similar to a barcode, that makes up the spectrum of light from a given object and, thereby is a signature of its chemical makeup, is shifted toward the red, or least energetic, end of the spectrum. One cause of red shift is the Doppler effect, which occurs when the object emitting light is moving away from the observer, thus having the effect of lengthening the wavelength of the light. Those astronomers already inclined to believe in Lemaitre's theory interpreted Hubble's observations of red shift as proof that the universe was expanding. Hubble, however, cautioned that redshift was not necessarily caused by recessional velocity. In 1941 he concluded that his observations did not support the expanding universe theory.

However, the Big Bang theory came to be the prevailing

theory for the origin of the universe among the scientific orthodoxy, and has remained so in spite of many problems squaring the theory with empirical data. According to the theory, the universe is about 15 billion years old. However, as telescopes have become more powerful and advanced, superclusters of galaxies have been discovered that could not have evolved in the time frame inherent in the theory. In other words, there are structures in the universe that are older than the age of the universe as postulated by the Big Bang theory.

Still, the scientific orthodoxy still clings to the theory. They keep making adjustments to it to make it work. They hypothesize concepts such as "dark matter" and "dark energy," and "negative pressure." The mathematical models of Big Bang scenarios are very arcane and complex, and require strict parameters to make them work. One is reminded of the Catholic Church's stubborn adherence to the geocentric cosmological model.

Why the stubborn orthodoxy? First of all, the Big Bang theory appeals to the same epistemological need as the origin theories of religion: the human psyche is hard-wired to seek out origins and first causes; we can conceptualize a timeline extending infinitely into the future,
but not into the past. The line does not have to end, but it had to have had a beginning at some point.

Secondly, many scientific careers and reputations have been built upon the Big Bang theory. Intellectual vanity among those in the scientific and academic orthodoxy has proven to be as formidable an obstacle as religious orthodoxy was to Galileo. Even the Enlightenment and Scientific Revolution could not allay the power of the human ego. But what of the alternative theories to the Big Bang?

In 1948, the astronomer and physicist Fred Hoyle and others formulated the Steady State theory, which asserts that the universe is eternal, and new matter is continually being created. Hoyle did not deny that galaxies could be moving away from one another, but likened this movement to water molecules in a river that moved away from one another yet left the river itself fundamentally unchanged. Hoyle considered the idea that the universe had a beginning as pseudoscience, as having no basis. But alas, the Big Bang theory won out with the scientific orthodoxy.

In the 1950's astronomers discovered celestial objects that were very bright but also very red-shifted. This meant that if red shift indicated that these objects were very far away, it followed that they would have to be very massive objects. Astronomers were puzzled as to how to classify them and named them "quasi-stellar objects," – quasars. But in the 1960's an astronomer named Halton Arp, who had once been an assistant to Edwin Hubble, noticed something strange: many high red-shifted quasars were not only in conjunction with low red-shifted galaxies, but appeared to be ejected from their cores. And thus was born a variation of the Steady state theory: the Continuous Creation theory.

According to Arps's theory the light from quasars is highly red-shifted because the atoms of which they are composed is new matter, and thus contain less energy than older matter. Thus, quasars are newly born galaxies. They are ejected from "active galactic nuclei," which are the dense cores of certain types of galaxies. After publishing his findings, Arp found that he could no longer get published in mainstream physics journals, and could no longer get telescope time at the big observatories. However, recognizing the groundbreaking nature of his discoveries, the Max Planck Institute for Astrophysics in Munich hired him. The scientific establishment had merely scoffed at Hoyle's theory, but they could not abide someone disproving the Big Bang theory with actual empirical observations. If quasars were newborn matter, this contradicted the entire premise of the Big Bang theory, that all the matter in the universe had been created in a single event fifteen billion years earlier. But how can matter be born?

We know that electrons can be created from light and vice versa. But what about protons and neutrons, which are part of a larger group of particles called hadrons? By definition, hadrons are particles that respond to the strong nuclear force, which is the force that holds atomic nuclei together. Aside from protons and neutrons, there are particles called mesons, which are less massive than protons and, some believe, mediate the strong force between protons; and hyperons, which are more massive than protons and neutrons, but are unstable, only exist for a small fraction of a second, and are only created in the controlled conditions of a laboratory.

The generally accepted mainstream theory is that hadrons are composed of yet smaller particles called quarks. When quarks were first proposed in 1964 the proponents of the theory suggested that they were not actual particles, but mathematical devices which enabled one to understand hadrons. However, the scientific community seized upon the theory and incorporated it, along with the Big Bang theory, into what became known as the Standard theory. In trying to prove the quark theory experimentally, scientists have to work around a tenet of the theory called "quark confinement," which asserts that quarks cannot exist freely and individually outside the atomic nucleus, but only exist as bound to other quarks. Thus, their existence has to be inferred indirectly. The experiments which have been done suffer the flaw of producing results that *could* support the quark theory, but could be explained in other ways as well.

An alternate theory – and one, of course, that did not get any traction with the scientific orthodoxy -- is the "space resonance" theory of Milo Wolff. He conceived of subatomic particles as spherical standing waves. By definition, a standing wave is two waves moving in opposite directions superimposed upon one another. He conceived of protons as having much higher density waves than electrons, likening them to waves inside a drum or a hollow sphere. He explained the strong force as the overlapping of these dense cores by two or more protons.

Consider how both the Continuous Creation theory and Space Resonance theory could be explained by assuming that the zero point field exists. Protons are created when electromagnetic waves are trapped in highly curved space inside the dense cores of active galactic nuclei. Once these cores reach a critical point they explode, spewing new matter into space which coalesces into new galaxies, beginning the process anew. We just need to adjust our perspective a bit, and everything becomes clear and simple.

But just what is this zero point field, this fabric of what we call space? Among other things, as we shall see, it is the well of all being, or rather, all the being of which we can conceive.

5

Western philosophy was born in the sixth century BC in the bustling Greek trading city of Miletus on the shore of what is now Turkey. Free thought and free trade seem to go together, and with the exchange of goods from all over the Mediterranean in Miletus came the exchange of ideas as well. The mathematician/astronomer/philosopher Thales of Miletus began the process of rational inquiry into the nature of the universe which blossomed into Western philosophy by speculating that water was the primordial element from which all the other elements were formed. However, his student Anaximander disagreed. He reasoned that the four elements air, earth, fire, and water had to have sprung from something other than themselves. This primordial source of being he called the *apeiron* – "the boundless."

In Anaximander's view, the *apeiron is* infinite in time and space and indestructible. He believed that there were many worlds, and that they originated when opposites such as hot and cold, wet and dry separated in the *apeiron.* This is an important concept, and one to which we will return: the universe we know and all its variety arises out of *nonuniformity* within an otherwise uniform, homogenous, isotropic medium. Aristotle later cited Anaximander when explaining the nature of space. In Aristotle's view, emptiness was not synonymous with nothingness. Because matter could occupy space, he reasoned, space had to have a real existence.

Philosophical speculations about the nature of space were few and far between after Aristotle, and with reason. In the pre-scientific age, when speculations about space were based purely on deductive logic, what could one surmise about space? Human perspective tells us that it is infinite, because we can conceive of nothing that could bound it. Our modes of perception lead us to the concept of "empty" space because objects can occupy it and move about in it, and to the convention of "dimension," which we associate with the ideas of volume and solidity.

But, as we have seen, thinkers started looking at space differently when it was demonstrated that light propagates as a wave, thus suggesting that space is either a medium or is permeated by one. This view was reinforced in the late nineteenth century when James Clerk Maxwell asserted that both gravity and electromagnetic forces require a medium in which to propagate. However, in 1887, the famous Michelson-Morley experiment, involving mirrors and a split beam of light that was designed to detect the "aether," the medium which pervaded space and allowed for the propagation of light, yielded negative results, discussion of the aether went by the wayside for a while.

But later, 1n 1911, Max Planck came to the startling conclusion that even if a portion of space were cooled to absolute zero and all matter and radiation were removed from it, that portion of space would still contain energy – zero point energy. Simply put, so-called "empty" space contains energy. In 1916, Walter Nernst proposed that space was "filled" with zero point electromagnetic radiation. A little later, in formulating his theories of gravity, Einstein asserted that space had intrinsic energy associated with it, that it had an "independent structure."

As physics progressed, empirical observations increasingly supported the existence of the zero point field. Even at absolute zero, helium would not solidify. According to the laws of physics, an electron orbiting a nucleus should lose energy and spiral into the nucleus. The fact that it did not indicated that it was absorbing energy from somewhere. Further, electrons were observed to emit and absorb photons spontaneously. Where was this energy coming from?

As quantum mechanics – the branch of physics that deals with subatomic particles – developed, the concept of "virtual particles" appeared. Simply put, this concept asserts that the energy within space can give rise to quantum particles spontaneously under the right conditions. In other words, *energy can become matter, and vice versa.*

Further, physicists observed other phenomena that suggest the existence of the zero point field. Particles moving through space create a flux analogous to the swell a boat makes in the water. Atoms and molecules exhibit jiggling movements that indicate they are being jostled by some energy source. And, in a phenomenon known as the Casimir effect, small electrically neutral plates placed parallel to one another will undergo an attractive force. If one assumes the existence of the zero point field, then the Casimir effect can be explained by the fact that there is much more energy outside the plates than between them, and this greater energy would tend to push them together. As we shall see, the Casimir effect may prove to be one means of extracting zero point energy from space.

So what is this zero point energy that makes up space? Is it some amorphous sea of energy, raw motion whose random machinations gave rise to what we conceive of as the universe? What is its nature? Can it be defined, diagrammed, delineated, captured in mathematical equations?

One clue to understanding the nature of the zero point field is the Planck length. As the name implies, this quantity was discovered by Max Planck. It keeps popping up in physics, and puzzles most physics. It is a very small length – 10^{-35} meters, which is *twenty powers of ten shorter than the diameter of a proton.* To conceive of how small that is, consider that the diameter of a proton is about twenty powers of ten shorter than the diameter of the earth. The reason the Planck length keeps popping up is that the forces of nature -- gravity, electromagnetism, and the strong and weak nuclear forces – seem not to have any meaning in spaces this small. At the Planck length the concept of space seems to break down.

But what if the Planck length is a wavelength? What if it

is the minimum possible wavelength of zero point energy? If this is the case, then one could understand why the forces of nature could not operate across a space smaller than the Planck length. The Planck length is the pixel of space, the very quantum of space. Using another analogy, it is like unto the mesh of a fishing net that can trap a minnow but not hold water.

So if the minimum wavelength of zero point energy is 10^{-35} meters, what is its speed? An axiom of relativity is that nothing can travel faster than light. However, calculations have shown that if the force of gravity did not update at many times the speed of light, then the solar system, let along the galaxy, could not hold together. Einstein dodged this inconvenient fact by asserting that gravity is not so much a force as the "geometry of space-time." However, there may be a grain of truth in this artful dodge.

The physicist Tom van Flandern, observing binary pulsars, calculated that the gravitational effect they had on one another updated at twenty billion times the speed of light. But this begs the question: what causes gravity? Einstein explained gravity in terms of
Riemannian geometry, which deals with curved surfaces. In Einstein's view, matter curved space. But what if space is flowing energy, zero point energy, and what if it is flowing at twenty billion times the speed of light? Then matter not so much curves space as diverts its flow. Therefore, gravity updates at twenty billion times the speed of light.

We know that it takes light about 8.3 minutes to travel to earth from the Sun, therefore the apparent position of the sun in the sky is not accurate. However, we know from experiments that the gravitational pull of the sun on the earth comes not from its apparent position but its real position. Orthodox physicists who take the axioms of relativity as holy writ either ignore this fact or twist themselves into pretzels trying to explain it.

But what of light? Why does it always travel at a constant velocity? Consider a row of dominoes. If they are of uniform size and weight and are spaced the same distance between each one, then no matter how much, or how little, force you use to knock over the first domino, then the rest will fall at a rate determined by the nature and spacing of the dominoes. And so it is with light. The speed of light is determined by the uniform nature of space. However, under certain special circumstances, light has been observed to travel a bit faster than its standard velocity. It is all a matter of altering the medium of space.

As Planck observed, light is emitted in discrete quanta – photons. These photons can, under the right conditions, be transmuted into electrons and vice versa. We know that protons are 1,834 times more massive than an electron. However, in the phenomenon of pair creation, a high energy photon gives rise to two particles – an electron and a positron. Consider the possibilities: in the dense conditions inside an active galactic nuclei, high energy photons get trapped, so to speak, in an area of high curvature, and, instead of radiating outward, they form stable spherical standing waves. If no pair creation takes place, it would take the energy of 917 photons to create this standing wave, this basic subatomic particle, this *proton.* Because electrons can be created by pair creation, and neutrons can be explained as composites of protons and electrons, then in this dense galactic core we have all the makings of hydrogen, the basic element from which all the heavier elements are synthesized inside stars. Once this core reaches critical mass it explodes, spewing newborn matter into space, matter that will become another galaxy of stars and planets and, presumably, life, hopefully intelligent life.

When we humans think of the universe, we think of stars and planets and galaxies, of light and atoms and molecules. This is our perspective, our *scale.* But let us consider the possibilities of the zero point scale, that dense sea of energy that gave rise to that portion of the universe we call home. The zero point field, what we call space, may be much more than a fount of raw energy. It may very well be home to vast, intricate energy configurations that, in terms of complexity, power, and *intelligence* put humanity to shame. Further, the infinitude of space may indeed comprise a holistic entity, a vast intelligence, a veritable Spinoza's God.

Let us consider the possibilities.

6

We began our inquiry with two questions: why is there something rather than nothing, and why are we here to ask the question? In a nutshell, the answer to the first question is that there is no such thing as nothing. It is an artifice of human epistemology and a mathematical abstraction. The answer to the second question could aptly be rephrased: how could something as complex as human intelligence exist? This question implies the premise that there is something problematic about the concept of order, of complexity. Indeed, the second law of thermodynamics asserts that energy systems tend toward ever greater levels of disorganization, of entropy. Organized complexity seems a great improbability, an anomaly – this in spite of the fact that we make that observation using our incredibly complex human brains. Oh, the wonder of irony!

But consider the snowflake. Beautiful in its hexagonal intricacy, no two alike, forming spontaneously from water vapor under the right conditions. And so too molecules and crystals, hexagonal and tetradecahedronal, epic art works of physics and chemistry. If we but look about at nature, and consider her machinations and transmutations, we arrive at the startling fact that

should not be so startling at all: *energy organizes itself.*

Consider the zero point field. Our hypothesis is that it is an infinite plenum of energy whose minimum wavelength is 10^{-35} meters and travels at 20 billion times the speed of light. Picture it as a uniform, homogenous, isotropic medium, unchanging in every direction forever. This is our starting point, our given. We will call this scenario A. Now, assume that some oscillator creates a wave in this zero point field, which has now become, by definition, a medium. Let us call this wave B. So now we have two entities, two conditions, two scenarios – A and B. Now suppose that B creates an unforeseen, unanticipated change in A, i.e., A +B = C. We now have three entities, three conditions, three scenarios. Now suppose that C + B creates yet another change, another configuration that reacts with A and creates D, and that D reacts with B and creates G which in turn reacts separately with A and creates H which reacts in different ways with all the other scenarios to create I, J, K, L, and M which interact in all their possible permutations in such a way that the alphabet is soon depleted. Thus, we arrive at a set of scenarios that were totally unpredicted by the initial conditions. And all of this arose out of *nonuniformity* in an otherwise uniform medium.

Of course, the assumption that at any time in the zero point field, indeed, in the universe – if the universe can be conceived of as being more or less than the zero point field – there was a condition of pure uniformity is an arbitrary assumption that proceeds from the nature of human epistemology. It is even self-contradictory in the sense that the essence of our model of the zero point field is motion, and motion by definition is change – to be is to move and to move is to be.

So let us consider the configurations that could arise in a medium of energy infinite in time and space. To do so requires that we move outside our perspective. When we conceive of life and intelligence, we think of our kind of life, our kind of intelligence. We think of carbon-based life. As we shall see, just the very presence of carbon in the universe suggests to some a fine-tuning of the constants of nature to allow such an element to exist. But what of life and intelligence that is not even based on carbon or any other element, but could arise within the zero point field itself?

Energy has the curious property not only of organizing itself, but of organizing itself into increasingly complex configurations. How order arises out of chaos has been a topic of speculation even before the advent of Thales and Western philosophy. The Greek myths speak of chaos as the primordial state of disorder out of which Cronos created the world. But what is disorder? Is there even such a thing? Is randomness disorder, or is there an inherent order behind randomness? In an effort to answer these questions, scientists and mathematicians have developed several related, interdisciplinary fields: *emergence, complexity, self-organization,* and *complex adaptive systems.*

Emergence, simply put, is the phenomenon in which an entity has properties its parts do not have – it is more than the sum of its parts. For example, carbon-based life is an emergent property of chemistry; a hurricane is an emergent property of humidity, temperature and convection currents; riots, and other manifestations of mass psychology, are emergent properties of the interactions of individuals under certain circumstances. Emergence is a holistic quality that is not only more than the sum of its pasts, but cannot even be predicted by the properties of its individual parts.

Closely related to emergence is complexity theory. Complexity theory focuses on the interactions and linkages between the constituent parts of an entity which yield the emergent whole. Complexity theory has applications in mathematics, computing, engineering, physics, and the social sciences.

Self-organization is the phenomenon whereby some form of order arises in an initially disordered system by an input of energy or a random fluctuation. Examples are phase transitions from solid to liquid to gas and vice versa; molecular self-assembly; and the spontaneous folding of proteins and other biomacromolecules.

A complex adaptive system is, as the name applies, a complex system that has the ability to mutate and self-organize in response to a change-initiating event. Living organisms are complex adaptive systems and intelligence is the highest, most complex characteristic of these systems. Imagine a complex adaptive system composed of zero point energy. Perhaps it could take the form of an intricate geometric manifold, circuits within circuits within circuits of energy flowing at 20 billion times the speed of light or more. Now imagine that this configuration was basically a grand brain with no limit in size and becoming bigger and more intricate every second, perhaps even permeating all of space – a cosmic intelligence. Or perhaps space is populated by myriads of such manifolds, creatures that to us would be as gods. It is not so hard to imagine. All we have to do is adjust our perspective.

The strong anthropic principle asserts that the constants of physics were fine-tuned to give rise to that portion of the universe which we humans call home. Before we consider the nature of the intelligence which could have done this fine-tuning, let us consider the merits of the strong anthropic principle.

7

"Would you not say to yourself, 'Some super-calculating intellect must have designed the properties of the carbon atom, otherwise the chance of my finding such an atom through the blind forces of nature would be utterly miniscule. A common sense interpretation of the facts suggests that a super intellect has monkeyed with physics, as well as chemistry and biology, and there are no blind forces worth speaking about in nature. The numbers one calculates from the facts seem to me so overwhelming as to put this conclusion almost beyond question.' "

*

"The reason why scientists like the "big bang" is because they are overshadowed by the Book of Genesis. It is deep within the psyche of most scientists to believe in the first page of Genesis."

Fred Hoyle

The astronomer/physicist Fred Hoyle (1915 – 2001) described himself as an atheist, yet he authored a book titled *The Intelligent Universe* in which he argued that the universe is "fine-tuned" to give rise to intelligent life. Is this a contradiction? It depends on how one defines atheism. If one believes that a super intelligence capable of bringing the atom-based scale of the universe with which we are familiar into existence is some transcendent being outside nature, outside the universe, then perhaps Hoyle's view is atheistic. If one believes that a "super intellect" could appear within the zero point field that had the power to bring forth a habitat hospitable to carbon-based intelligent life, then the charge of atheism would depend on how one defines the concept of God.

Hoyle, it will be remembered, scoffed at the idea of the Big Bang. Indeed, he coined the very term as a pejorative for the idea that the universe popped out of nothingness a mere 15 billion years ago. Unlike too many physicists and astronomers, Hoyle had the ability to consider the big questions of cosmology without abandoning common sense. Something cannot originate from nothing. The universe, by definition all that exists, must have always existed. He saw the Big Bang cosmologists as being hobbled by that age-old conundrum of first cause, and their stubborn adherence to the theory as assuaging this epistemological need. However, when his calculations based upon careful empirical observations suggested to him that some intelligent force had been involved in the fine-tuning of the forces of nature, he did not hesitate to state as such.

Ironically, many of the same physicists who ridiculed Hoyle's Steady State theory also scoffed at his idea of a cosmic intelligence. Consider that: adherents to a cosmological theory that posited a specific creation event rejected the idea that a super intellect could fine tune
the constants of physics. Hoyle managed to avoid this logical contradiction by simply looking at the data with fresh eyes, using his intellect to see beyond the limits of human perspective.

What led Hoyle to his idea of cosmic fine tuning, of the anthropic principle, is his theory of nucleosynthesis, the process whereby elements heavier than helium are formed in the dense cores of stars, which then spew them into space when the explode as supernovas. Hoyle introduced the idea of nucleosynthesis in a research paper in 1946. In this paper he explained that nucleosynthesis occurs not as a result of random collisions of atomic nuclei, but follow specific pathways determined by the energy levels of the nuclei of specific elements. For example, owing to what he called the "e process," Hoyle demonstrated that iron is much more abundant than other heavy elements owing to thermal equilibrium between nuclear particles. However, in a second paper published in 1954, he showed that the elements on the periodic table between carbon and iron cannot be synthesized from such equilibrium processes. He concluded that the "triple-alpha process" which generates carbon from helium would require the carbon nucleus to have a very specific energy level and symmetry for it to work. Because this specific energy level is very unlikely to appear in nature, Hoyle was puzzled by the great abundance of carbon in the universe.

Carbon is a unique element whose unusual chemistry allows it to bond to itself and other elements to create highly complex organic molecules that are essential to life. According to Hoyle's calculations, which were later borne out by experiment, there should be very little carbon in the universe. But against all odds, it is quite abundant. Further, Hoyle turned his attention to other improbable cosmic "coincidences." For example, he compared the odds of a single functioning protein molecule forming by random chemical processes to that of a star system of blind men simultaneously solving Rubik's cube.

Following Hoyle's lead, other physicists calculated other incredible coincidences. For example, if the force of gravity were just a little weaker, stars would not form from hydrogen gas. If the strong nuclear force were just a little stronger, elements heavier than hydrogen would not form. If the electrical charge of the proton and electron were just a little different, atoms would not form In addition to Hoyle's book *The Intelligent Universe,* the strong anthropic principle is discussed at length in *Cosmic Coincidences* by John Gribbin and Martin Rees.

Critics of the idea of intelligent fine-tuning assert the possibility that life could be based upon other elements. This idea is indeed very possible, but it in no way negates the fact that the abundance of carbon in the universe is indeed a very unlikely coincidence. Further, just as Hoyle saw the odds of spontaneous formation of proteins as mathematically improbable, many observers point to that most complex of all organic molecules –

DNA – as strong evidence of the guiding hand of a super intellect in the appearance of life.

Deoxyribonucleic acid is composed of chemical compounds called bases that bind together in pairs. Adenine always binds with thymine, and cytosine always binds with guanine. Each half of the base pair is held in place on a long strand of the sugar deoxyribose. Thus, two strands of deoxyribose together form a double helix. Groups of three bases code for specific amino acids which in turn comprise proteins. In the human genome, there are 3.2 billion base pairs. If we represented each base pair by just one character in small print such that each page had sixty lines of a hundred characters each, it would take over five hundred volumes of a thousand pages each to represent the human genome. Admittedly, this is not proof of intelligent design by mathematical certainty, but would it not rise to the level of a reasonable assumption?

Hoyle believed that life appeared on earth by means of "panspermia," the process by which organic compounds drift through space, or are perhaps borne by comets and meteors, and come to rest on suitable planets and seed life. He rejected the idea held by mainstream Darwinists that life arose on the early earth in the "primordial soup" where, supposedly, hydrogen compounds combined to form amino acids which in turn gave rise to proteins and other organic compounds.

The idea of panspermia was not original with Hoyle. Earlier thinkers had suggested the idea of "directed panspermia," that is, the idea that advanced civilizations had deliberately seeded Earth with life. As we have seen, John von Neumann proposed a very similar idea. As we shall see, there may be some merit to the idea of directed panspermia. However, it begs many questions. How did carbon based life originate? If the constants of our anthropic universe were fine-tuned to give rise to life, who or what fine-tuned them? If we are the creations of a super intellect, then what is the nature of this super intellect? What was its motive in creating humans? Can we deduce things about this super intellect just by the very fact that we have the capability of even conceiving of such a being?

    To consider these possibilities, let us return to the zero point field, that place where infinite energy together with infinite time and space yields infinite possibilities.

8

In the short story "Crusades" by Arthur C. Clarke, the protagonist is a planet whose main constituent element is helium. This planet has drifted into interstellar space where the temperature is just barely above absolute zero. At such a low temperature helium becomes a liquid, and, as such, becomes a perfect superconductor in which electricity flows at nearly the speed of light. Over the eons this superconducting sphere evolves into a huge brain, a superintelligence. Realizing that eventually the tug of gravity will pull it toward one of its neighboring galaxies and, thus, into a milieu too hot for helium to exist in liquid form, the brain uses certain elements available to it to create space probes to explore the neighboring galaxies and transmit back its findings. Eventually, the probes discover carbon-based life forms that exist on planets which, by the standards of the mother planet, are incredibly hot. Further, the probes discover that these supple, watery carbon-based life forms utilize silicone-based devices they consider artificial intelligence – computers. Thus, the probes arrive at the conclusion that carbon-based life has enslaved their kind of life. When they report this back to the mother planet/brain, it orders them to destroy carbon-based life wherever they find it, and the crusades begin.

Clarke's short story is remarkable for several reasons. First of all, using sound scientific principles, it imagines a life form totally different from us. It does so using the underlying principles of complexity, emergence, self-organization, and complex adaptive systems. The planet/brain is an example of the ultimate emergent property – intelligence, a force that perceives, reasons, imagines, creates, and takes action. In this instance the intelligence is holistic in that it is comprised of all its constituent matter and the configurations thereof into a unitary entity.

Clarke's planet/brain emerged from liquid helium because that medium greatly facilitated the flow of energy in a sustained configuration. Imagine, then, the possibilities within the zero point field where energy is many, many orders of magnitude greater than anything on our scale. Imagine the configurations which could emerge that were not only self-sustaining, not only constantly growing and evolving, but doing so in a conscious, self-directed manner. Complexity begetting greater complexity, so on and so on *ad infinitum.*

Perhaps the key to understanding the possibilities of zero point beings is geometry. What configuration would a flow of zero point energy assume that was self-sustaining, constantly adapting, and had the qualities associated with higher intelligence – curiosity, imagination, creativity, and self-aggrandizement? Perhaps such a configuration could be expressed in terms of fractals, geometric shapes composed of smaller versions of itself. Or perhaps a new branch of geometry would have to be invented, one that incorporated the principles of crystallization and fractals with the idea of constant motion and change, of emergence and complexity.

What would the values of a complex zero point intelligence be? First of all, as for all creatures, it would be self-preservation. Without this drive, no living creature would long endure. And self-preservation is perhaps the basis of all other instincts and faculties of living beings. Curiosity, imagination, creativity, self-esteem, aesthetics – all of these are ultimately based upon the logical, powerful urge of an intelligent life form to perpetuate its existence. However, on the zero point scale, self-preservation may not be such constant struggle as it is on our scale. Here on Earth, and presumably on other planets where carbon-based life is the norm, constant effort is required to secure the necessities and amenities of life, and to protect ourselves against the dangers and vicissitudes inherent in our existence. Zero-point creatures, being composed of and existing in a milieu of readily available energy and infinite, unbounded space, would perhaps not see life as a struggle, but as something far different.

A zero point being would be a creature of great intelligence. But what is intelligence? First of all it is the ability to reason. Reason can be defined as the irreducible laws governing reality. A thing either exists or it does not; something cannot originate from nothing; natural phenomena are subject to the laws of cause and effect; the laws of nature are immutable, and to ignore them invites peril. Thus, an intelligent being is, by definition, one that considers its situation and actions logically as opposed to acting rashly, blindly.

If reason is the foundation of intelligence, then curiosity, imagination, and creativity form the super structure. Before Thomas Edison invented the phonograph and motion pictures, he had to imagine, first of all, that such things could exist, and secondly, how to make them a reality. From childhood, Edison had been endlessly curious about what makes things work. It was from this curiosity that his imagination and creativity grew.

Consider the wonder of imagination. It is the ability to conceive of that which is not. Consider the miracle of creativity. It is the ability to bring the imagined into reality. Both these qualities are emergent properties of intelligence. Thus, consider the great, restless, irrepressible urges a complex zero point intelligence would have to imagine and create. Further, consider the power and nature of such a being, having infinite energy at its disposal and the ability to wield and direct this energy. Such power would be, in our lexicon, godlike.

We humans have had an obsession with the idea of gods, of creatures more powerful than ourselves who dwell in the heavens, even before the beginning of civilization. Perhaps the idea of such beings is part of a deep, atavistic memory embedded in our DNA. The Greek myths as well as Hebrew scriptures speak of celestial creatures who bedeviled, guided, and interbred with humans. The idea that there are other intelligent creatures in the universe and that they interact with man seemed commonsensical to our ancestors. Perhaps they were not so far off the mark.

The universe may very well be teeming with life forms we cannot begin to imagine. Further, there may even be life forms that exist on a scale even smaller than the Planck scale. With our limited perspective, the Planck scale is the smallest we can conceive of. The zero point field may itself be the artifice, the creation, of an intelligence that exists on a scale that the human mind would find difficult, if not impossible, to fathom. But the very fact that we can even conceive of such a possibility suggests something about our nature as humans as well as the universe itself: intelligence is not just a byproduct of the universe, but its very source, its very purpose, its very essence. In terms of Aristotelian causes, intelligence is both the efficient and final causes of the universe.

In order to understand how limited our perspective is, consider Halton Arp's observation that the only part of the cosmos we can be sure that we are observing is the Virgo and Fornax superclusters. If the universe is infinite, and the carbon-based macro scale of stars and galaxies is
limited to two superclusters, then our scale of existence is just a mote of dust in the universe overall. The variety of life forms possible in the infinitude of space is just that – infinite. Perhaps we humans could best understand the nature of our existence if we conceived of ourselves as a mode of being, as just one form of intelligence. In the eternity of time, sparks of intelligence may take different forms, different modes. Perhaps humans, in their apparent mortality, are just a means by which pre-existing intelligence is reborn and discovers the wonders of the universe anew time and time again.

Let us consider the possibilities.

9

What a piece of work is a man! How noble in reason! How infinite in faculty! In form and moving express and admirable! In action how like an angel! In apprehension how like a god!

Hamlet, 2:2   -- William Shakespeare

The Earth is 3.5 billion years old. The oldest fossils of single cell organisms are about 3 billion years old. The simplest one-celled bacteria has about one million base pairs in its DNA. If we represented each base pair by a single character on closely printed pages of one hundred characters per line and sixty lines per page, it would take over 160 pages to delineate the genome of the simplest bacteria.

Life began in the sea. Land life did not appear until about 600 million years ago. Mammals appeared about 50 million years ago, and the first humans about 300,000 years ago. If the entire 3.5 billion years of the Earth's existence were represented on a timeline a mile long,

the 5,500 years of civilization would take up a bar at the tip of that line about a sixteenth of an inch thick. The line representing the two hundred and fifty years since the beginning of the Industrial Revolution would be microscopic. In that flash of time mankind has gone from steam engines to nuclear reactors, from horse-drawn carriages to spacecraft. What are we to make of that? What can we deduce about the universe just by the very fact of our existence?

Taken as a whole, life on earth seems to be a process, an unfolding. When the Earth first formed, the main constituents of the atmosphere were hydrogen and hydrogen compounds. Oxygen did not appear until the appearance of photosynthetic organisms that gave off oxygen as a byproduct of metabolism. This oxygen shield blocked harmful radiation that would have precluded the development of advanced life forms. Everything was in sync to bring forth an ecosystem favorable to the appearance and perpetuation of an intelligent species. Further, consider the fact that there is only one intelligent species on Earth.

Human life is limited. We do not comprehend the great unfolding of which we are a part, nor probably are we supposed to comprehend it. Our lot is to further the unfolding just by following our innate strivings and compulsions. This collective, multigenerational effort seems to be moving toward the goal of mastering the forces of nature and traveling to the stars. That goal seems to be hardwired into our collective DNA. But are we merely mindless automatons serving the purposes of higher powers we can infer but not comprehend? Or are we more than that?

Attributes of intelligence and physicality are distributed unequally among humans. The bell curve is an inescapable fact of our existence. Were it not so, there would be no functional society. If everyone's IQ was on the left side of the curve, we would probably still be in the
stone age. If everyone were of high IQ, then there would have been no one to perform the onerous tasks which built civilization and all the wonderful products thereof. And for most of human history, life was just that for the vast majority of people – onerous.

As Thomas Hobbes observed in the 17 th century, "for most of mankind life is short, harsh, and brutish." The lives we live today would seem incredibly desirable to most people born just a mere century ago. But in spite of our vastly increased leisure time and access to an ever increasing fund of knowledge, relatively few people delve deep into the mysteries of life and the universe. They just know they are here, stuck in the human predicament, and are trying to make the best of it.

But consider the gnawing frustration of intellectually curious people throughout the ages. We are born into this world without a clue as to how mankind and the world came into existence. As Nikos Kazantzakis wrote, "life is a stream whose source is hidden." Born in the late 1800's, Kazantzakis was a restlessly intelligent man who searched far and wide for the answer to the riddle of life. He came to the conclusion that there was a God of "mighty power", but not an omniscient, omnipotent God. Seeing evidence of design in the universe, he rejected the purely materialist, Darwinian view that life arose by some sort of random accident. However, he found the idea of God as some sort of metaphysical abstraction untenable as well. Samuel Butler also rejected the idea of God as an abstraction. In his book *God the Known and Unknown,* Butler put forth the proposition that God is a distinct individual as opposed to some holistic abstraction.

However mankind came to be, there are some inescapable facts about our nature. Perhaps we can best be described as robots made of carbon-based compounds whose nature it is to use our incredibly complex brains to survive, thrive, advance, and ultimately to find a way to travel to other planets when this one becomes unlivable. In other words, we may be von Neumann probes. Further, consider the fact our planet abounds in all the materials we need to accomplish this destiny. That is a strong clue to our nature and origins. And the beauty of our human predicament is that all each of us has to do is follow our own individual desires and inclinations, and the collective interactions we have with each other results in a society capable of transporting its members to the stars.

The lifespan of an individual human is excruciatingly limited. Therefore, humans specialize in their means of earning a livelihood. An individual cannot produce all the goods he needs and desires, and therefore trades with other human beings. "No man is an island, entire of itself," as John Donne so cogently observed. And this is part of the human predicament. We have the intellect to imagine all the possibilities of life, but are limited by our material circumstances. Before he committed suicide, the Japanese writer Yukio Mishima wrote: "Human life is limited. I would like to live a thousand years." A byproduct of intelligence is the ever-present consciousness of our mortality. To assuage this existential malaise, man, early on, developed varying notions of an afterlife. Plato believed in the transmigration of souls – reincarnation. This was his way of making sense of human life, which is ennobled by such grand possibilities but so limited by time and circumstances.

Perhaps the best way to assuage the befuddlement we all feel when we contemplate the vexatious nature of our existence is to consider the great mystery inherent in our very existence. Perhaps there is more to life than we understand, perhaps not. But the awesome mystery remains. In the twenty-first century, we are many orders of magnitude closer to understanding the mystery than we were just a bit over a century ago. But consider what it must have been like for the earliest humans.

Our best current estimate is that the first hominids that could truly be described as *homo sapiens* appeared about 300,000 years ago. If the 3.5 billion years of the earth's existence were represented on a timeline a mile long, 300,000 years would take up three inches at the end of the line. For the great majority of that three hundred millennia, the new species were hunter-gatherers. It was only about ten thousand years ago that our progenitors cultivated agricultural crops. It was only about 5,500 years ago that they settled down in one place long enough and had enough surplus foodstuffs that civilization was able to develop.

Consider the predicament of the intellectually curious among early man. No philosophy or science or mathematics to fall back on, they must have been truly befuddled by their very existence. Reasoning that some higher power had brought them into being, the idea of gods evolved in human society.

Evolve. It is ironic that the Darwinists use this word. The word is derived from the Latin word for "to unfold." Unfolding implies a process toward a predetermined goal, such as the germination of a seed which gives rise to a plant or a tree. From our vantage point the history not only of mankind, but of life itself seems like an unfolding towards an end. However, for most of human history the basic conditions of human life were static. Most people worked in agriculture, society was tightly stratified, land transportation was effected by animal power and seafaring by sail, and the earth was thought to be the center of the universe. Even the great Greek philosophers never imagined the great scientific and technological advances that would essentially and intrinsically change humankind and, in the process, suggest to them that they were part of a great unfolding.

The German philosopher G.W.F. Hegel, born in the early years of the Industrial
Revolution, in trying to make sense of human existence, had to deal not only with the phenomenon of change in society, but of progress. Versed in history, he was well acquainted with the cycles of civilization, of how they rose and fell. But the Enlightenment, the Scientific Revolution, and The Industrial Revolution, all of which he was living in, presented him with a conundrum: history was not just a series of cycles, it was a spiral, an unfolding. And in complex reasoning that is often inscrutable, he sought to explain the nature and goal of this unfolding.

Hegel's system prefigured emergence theory. His view of nature and human history can best be summed up in a line from William Blake's "Marriage of Heaven and Hell": "Without contraries is there no progression." He called his method "dialectic," which is to say that it involves the interplay of opposites in the same way that a debate between men holding opposite views may result in some middle-ground truth. Hegel saw history as an unfolding of the *Weltgeist,* which can be interpreted as either "World Spirit" or "World Mind" into its "otherness" – nature. He saw the pinnacle of this unfolding as an orderly, free, enlightened society where men were aware not only of the unfolding but the purpose thereof. In this way, the *Weltgeist,* by existing in a form outside itself, arrived at "absolute consciousness" of itself.

Much of Hegel's concepts are vague, to say the least. He was trying to express concepts that , from his limited 18 th and 19 th century perspective, were inexpressible. What did he mean by *Weltgeist?* Did he mean an abstract Platonic "form", or did he mean a holistic being such as Spinoza's god? He never really clarified what he meant, perhaps because he could not. In his *History of Western Philosophy,* Bertrand Russel observes that almost all of Hegel's ideas have been proven false. Furthermore, Hegel's philosophy is anathema to some because Karl Marx based his system of "dialectical materialism" on it, and to still others who bemoan the fact
that Hegel, in his last book, glorified the militaristic Prussian state. In spite of all this, Hegel's system deserves study for at least one reason: he discerned that human society was not only changing but progressing, and he tried to discern what this progress was leading to.

It is too bad that Hegel did not live in our times. If he thought the age in which he lived was marked by unprecedented change, imagine what he would have thought of the twentieth century. In 1900, the lifestyle of most people was little changed from what it had been in 1830, the year of Hegel's passing. The most significant changes in that seventy years had been the advent of the railroads and the telegraph. But things were about to change drastically with the advent of Edison, Ford, the Wright brothers, Marconi, Planck, Einstein, and an army of thinkers and inventors inspired and enabled by these giants. In less than fifty years from 1900, not only were the roadways filled with automobiles and the skies with airplanes, mankind had done something that no one had ever dreamed of in times past – split the atom. Twelve years after that, the first satellite was launched into orbit, and not long after that humans were traveling into space. What would Hegel had made of all that, and all the wondrous achievements to come after? Would his notions of the *Weltgeist* be the same, or would he have a more concrete idea of the cosmic intelligence that lay behind the mystery of humanity?

What would he see as the destiny of mankind?

We live in a great age, an age that the great thinkers of yore would have loved to have lived in had they but been able to imagine it. So let us attempt to do what they could not – imagine what is to come.

10

In that unprecedented, incredible explosion of scientific and technological advancement that began in the early 1900's, perhaps mankind's very psyche changed. The human brain, suddenly and without warning, was confronted with possibilities that went beyond its hardwired preoccupations with survival and reproduction. Our purpose – and our origins – were implied by the fact that not only had we unleashed the energy congealed in atomic nuclei, we were venturing into space. The veil had been lifted a bit. We were forced to seek answers to questions which heretofore no one had thought to ask, or to at least take seriously.

In retrospect, our foray into the heavens took place rather abruptly. The space age began in October, 1957 when the Soviets launched Sputnik, the first artificial satellite into orbit. The idea of communication satellites had been proposed years before by the science fiction writer Arthur C. Clarke. The rush to put manned space probes into space arose out of our military and political rivalry with the Soviet Union. Hegel would have recognized the space age as a product of the dialectical clash between two opposites, just as the earlier race to split the atom had arisen between America and Nazi Germany – progression from contraries.

Although in his later years, probably out of political expediency, Hegel had extolled Prussia as the ideal state in which the *Weltgeist* was finally embodied, he had earlier opined that "America is the land of the future." The way things unfolded, his earlier prediction proved correct.

Humankind's foray into space, though an existential

milestone in our unfolding as a species, raised many knotty questions. We were in space, but for what purpose? Some visionaries began to dream of interstellar travel. But right away Einstein's dictum that nothing could travel faster than light led many to doubt the feasibility, if not the very possibility, of that dream. In answer to these naysayers, some thinkers came up with exotic ideas like "space-time warps" and "wormholes" through space. Oddly, very few people questioned Einstein's cosmic speed limit. Yet we continue to probe space. Voyager, an unmanned space probe, just a few years back left the solar system, venturing into interstellar space, and continues to send data to Earth.

It is as though something in the human DNA draws us toward the heavens. And with reason. The cosmos is an extremely precarious place. An unexpected solar anomaly or a killer asteroid could wipe out humanity at any time. However, if we are to travel to and settle Earth-like planets in other solar systems, the distances involved demand that we learn how to travel at speeds faster than light. Is such a thing possible? Perhaps the answer to that question lies not so much in contemplation of our future, but of our origins.

If the strong anthropic principle is correct, then stars and planets were designed to give rise to intelligent life. By force of gravity, stars congeal from nebulae of hydrogen gas. In the hot interiors of stars heavy elements form as the nuclei of lighter elements fuse. As some stars reach
a certain critical mass, they explode as supernovae, thus spewing heavy elements into space, which then form accretion disks around other stars and gradually coalesce into planets. When conditions are just right, a planet like Earth forms whose conditions are optimal not only for the emergence of life, but of intelligent life.

Consider Earth. Not all that deep in the crust of our planet we humans have all the raw materials we need not only to sustain life, but to create an increasingly complex technological society capable of launching humans into space. If all of this is the product of pure happenstance, it is one incredible sequence of coincidences. If not, then the question remains of how life arose on Earth.

If we are the product of a superintelligence that exists within the zero point field, then one possibility is that this intelligence engineered Earth-bound life from scratch, and that this life unfolded from the first single-cell organisms that first appeared 3.5 billion years ago. Another possibility is that fully developed *homo sapiens* either traveled here – or were placed here – by an advanced civilization that had achieved faster-than-light travel after the lower life forms had evolved. Yet another possibility is that after *homo sapiens* evolved here on earth, travelers from advanced civilizations came here to guide us and help us in our own unfolding as a species.

Our myths, folklore, and religion are rife with tales of beings who descended from the heavens to give direction and succor to our ancient progenitors. If we accept the premise that interstellar distances are not that daunting an obstacle to a sufficiently advanced civilization, then the idea of extraterrestrial visitors to Earth does not seem farfetched at all. The question then becomes how to achieve faster-than-light travel. Before we consider that question, we have to consider one more basic: how do we extract and use zero point energy.

The irony of zero point energy is that although we are composed of zero pointy energy and exist in the medium of zero point energy, the scale on which we exist makes it not only difficult to extract zero point energy, but even to imagine how to do it. And there is a very good reason for this. We humans are of such a nature that our technological advancement often outstrips our political and cultural development. Achieving the ability to split the atom brought us to the brink of nuclear war. Needless to say, the quest to extract zero point energy must proceed with great caution and deliberation. And perhaps that is the role of America. Although we still have much work to do in setting our own house in order, we are still the world's preeminent superpower. Before we learn how to extract and utilize zero point energy, we have to give great thought to what we are and where we are going as a nation.

But can zero point energy extraction be done? Actually, there are those who are already working on it.

In 2008 a patent was granted for a device that uses the Casimir effect to extract zero point energy by manipulating the electron orbits around hydrogen atoms. The Casimir effect, it will be remembered, is the phenomenon whereby two electrically neutral metal plates placed close to one another will experience an attractive force due to the fact that there is more zero point energy outside the plates than between. The "quantum energy vacuum energy extraction" device invented and patented by Bernard Haisch and Garnet Moddel involves pumping hydrogen gas through tiny chambers so small that certain electron orbits are not allowed, and thus the hydrogen atoms give up energy, but regain it upon exiting the chamber by absorbing energy from the zero point field. The energy thus trapped can, theoretically, be used to generate heat and
electricity. The device, of course, is not commercially viable yet, but it demonstrates at least one pathway to extracting zero point energy.

Perhaps another possibility is using nuclear energy or lasers to concentrate a great amount of energy on a tiny portion of space, thus creating a "tear" or an area of great non-uniformity in space. Just as, according to the continuous creation theory, new matter is created in the high energy dense cores of active galactic nuclei, perhaps a smaller area of concentrated energy could release more zero point energy than was expended to release it. In any event, it seems that tapping into zero point energy is either our hard-wired objective as a species, or should become our chosen purpose if we want our species to survive. Not only will Earth not last forever, any number of extinction level catastrophes could befall us at any time.

But if we survive as a species long enough to enter the zero point age, a truly incredible age it will be. Let us consider the possibilities.

11

An unlimited supply of incredibly cheap, clean energy: what will this mean for the world? An explosion of economic activity never seen before; a much cleaner planet; and certain political and economic realignments when nations that depend on oil exports for their wealth and power see their main resource plummet in value, and nations that possess the ability to extract zero point energy see their wealth and power soar. But there is a downside to zero point energy. In the wrong hands it could power a doomsday weapon. Thus, to put it bluntly, one nation must have a monopoly on the technology for zero point extraction and utilization. And it is clear that that nation must be the United States.

If history teaches anything it is that there will never be world peace until one nation has an overwhelming monopoly on military might. Rome proved this during the two hundred years of the *Pax Romana.* If not for the sake of lasting world peace then for our own national security the United States must develop and maintain a monopoly on zero point energy. However, this may not prove to be an unpopular state of affairs. Peace gives rise to prosperity. People whose lives are worth living are not so easily coaxed into aggressive wars by ambitious despots. Cheap, unlimited energy will be a boon to developing nations looking to catch up with the West.

Cultures are not equal. Imagine what the world would be like if the United States were not strong enough to put a check on Russia, China, and Iran. Due to wrongheaded policies the superpower status of the U.S. has already been eroded somewhat.

Russia, China, and Iran are already testing us, and are forming alliances with one another. Thus, once we truly enter the zero point age, certain political realities must be faced. And once they are, the world will truly see a golden age: the *Pax Americana.*

Aside from the economic and political changes wrought by the zero point age, the scientific and technological advances made possible by utilizing zero point energy will be the fulfilment of our very purpose as a species, will bring about a fundamental existential change to Earth-bound humanity. Consider: *harnessing zero point energy will allow us to communicate, compute, and travel at twenty billion times the speed of light or more.* Let us consider the possibilities.

SETI – Search for Extraterrestrial Intelligence. We have all heard of this organization. They utilize radio telescopes in various places around the globe in an effort to pick up electromagnetic transmissions from advanced civilizations on other planets. Skeptics say they have never succeeded. However, the plain facts indicate that they have had one success.

On the evening of August 15, 1977 the Big Ear Telescope at the University of Ohio intercepted an electromagnetic transmission that originated in the vicinity of the star Tau Sagittarii that had all the earmarks of a transmission from an advanced civilization. First of all, the signal was much too strong to have been generated by anything on Earth or in Earth's orbit. Further, the frequency of the signal was 1.42 gigahertz, the frequency of hydrogen in its excited state. Searchers for extraterrestrial intelligence had long speculated that an intelligent species reaching out to another intelligent species would transmit on a frequency the significance of which would be recognized. Hydrogen is the basic element in the universe, and 1.42 gigahertz was a prime candidate for a possible alien signal. Finally, for years the skeptics searched for an alternative explanation for the signal -- dubbed the Wow! Signal -- and found none.

The Wow! Signal lasted seventy-two seconds – exactly the

time a signal from a fixed point in space lasts when arriving at the rotating Earth. The frequency nor the amplitude of the wave was not modulated, therefore it did not carry a message. Given its great strength, the transmission was more in the spirit of a beacon announcing an intelligent species' presence.

Years ago, a science writer for a popular science magazine mused about the reasons for the lack of messages from alien civilizations. He arrived at the conclusion that in all the universe perhaps we were the only planet whose conditions had been just right for the origin and evolution of intelligent life. Just on the face of it this argument seems preposterous. Just in our own galaxy there are about 200 billion stars, and there are untold billions of galaxies in the known universe. A more logical explanation is that electromagnetic signals degrade over the vast distances of space, making them too feeble for detection by our technology. Tau Sagittarii is relatively close to Earth – 120 light years, and the Wow! Signal was incredibly strong, indicating a source capable of generating a great amount of energy.

But perhaps there is an even simpler reason. Super-advanced civilizations, knowing the limitations of electromagnetic transmissions, and having the capability of tapping zero point energy, are transmitting with zero point energy. Not having reached zero point capability yet, we cannot receive their transmissions. Consider the advantages of using zero point energy for communication. First of all it would be twenty billion times the speed of light or more. Further, traveling through the zero point field the signal would not lose energy and the signal would not degrade. With zero point energy we could communicate with Tau Sagittarii, for example, in a tiny fraction of a second, whereas the Wow! signal took 120 years to reach us.

The possibilities of zero point communication are stunning. The Milky Way galaxy is 100,000 light years across. A zero point signal could traverse that distance in a little over two minutes. Further, remember that we are assuming that 20 billion x c is the maximum speed of zero point energy because that is the observed velocity of gravity. It could possibly be manipulated by an advanced civilization into moving even faster. Thus, our galaxy and universe may not be as large – and as lonely – as we once thought.

But how do we utilize zero point energy for communication, computation, and travel? Even when we learn how to tap into zero point energy and power all our various and sundry devices and machines, how do we then modulate and manipulate zero point energy, channel its flow, and make it conform to our purposes? In order to answer these questions, we must first redefine and reappraise some of very basic modes of perception, chief among them motion itself.

We think we understand what motion is. An object moves from point A to point B and that is motion. The motion of a projectile, be it a cannonball or a baseball, can be described very precisely in terms of a parabola. But what about the motion of a single atom? In the early twentieth century very gifted physicists like Max Planck, Werner Heisenberg, Wolfgang Pauli, and Louis de Broglie discovered that the motion of atoms and subatomic particles could not be described precisely in the same way that the motion of macro objects can be. In a space between point A and point B, a quantum particle in motion could be in any number of positions due to its erratic trajectory. Quantum particles are subject to incessant jostling from zero point energy, and thus do not travel in a precise trajectory. The position of quantum particles can only be described in terms of probability. On a graph, the amplitude, that is, the height, of a probability wave derived from a trigonometric function, represents the probability of finding a quantum particle in a given position. In short, the quantum scale operates much differently from the macro scale.

Perhaps the most vexing prediction that arose from the mathematics of the then new science of quantum mechanics was the idea of "quantum entanglement." Certain equations suggested that when two quantum particles interacted, their states became interdependent – "entangled" – and remained entangled even when separated by vast distances. No matter how vast the distance a change in the state of one particle would induce an instantaneous corresponding change in the other, thus violating Einstein's dictum that nothing can travel faster than light. Derisively, Einstein called this idea "spooky action at a distance," and dismissed it as impossible. However, later experiments, some as recently as the year of this writing, proved that faster than light interaction between entangled particles is indeed an actual phenomenon.

What are we to make of this? First of all it would seem to confirm zero point theory, particularly that zero point energy travels much faster than light. Further, it leads us to wonder about the nature of particles, and whether our notions about them need to be revised. Does the entanglement between two particles form some sort of bridge or conduit through the zero point field, thus forming one composite entity instead of two separate ones? Again, we have to reexamine our ingrained habits of perception.

The key takeaway from quantum entanglement is that the flow of zero point energy can be modulated. Thus, zero point communication is possible. But if using quantum entanglement for communication is still in the visionary stage, using it for computing is already in the experimental stage. In 1985 the British physicist David Deutsch outlined how a computer based on the principles of quantum physics as opposed to classical physics would work. Instead of using base 1 binary code – ones and zeroes – to represent bits of information that correspond to on and off positions of electronic switches, quantum computers would use electron spin that would allow quantum bits – qubits – to be on, off, and something in between simultaneously. For example, using just forty qubits could enable a quantum computer to process over a trillion numbers simultaneously. Quantum entanglement comes into play by eliminating outside electrical interference that would disrupt the electron spin.

To date, quantum computers have indeed been built, but are still impractical for commercial use. However, progress is being made. But quantum computing is not zero point computing, does not operate at faster than light speed. But consider how achieving the capability of zero point communication could lead to zero point computation. On/off circuits opening and closing at twenty billion times the speed of light or more would give rise to computers whose power we could scarcely imagine. Computational power of this magnitude could perhaps bring us into fellowship with that inscrutable cosmic intelligence that engineered us in the first place.

But what of zero point travel. Will we ever travel at twenty billion times the speed of light or faster?

Let us consider the possibilities.

12

What is motion? Like many concepts – space, time, energy, matter, universe – physicists treat it as an undefined term seemingly self-evident in its nature. Be it a macro object like a planet or a baseball or a quantum particle, mathematics can *describe* its translation from one point to another, but, at its most fundamental level cannot tell us just what this translation is. We know that a macro object is a composite object made of atoms, and, thus, the movement of a macro object is the collective movement of all its constituent atoms. But what of the movement of atoms and subatomic particles? Do they move in the same way that we *think* macro objects move?

Consider the phenomenon of quantum tunneling, in which atoms and subatomic particles pass through barriers of matter or energy which, presumably, they do not have enough energy to do so. Physicists have known about this phenomenon since the 1920's, and explain it by the wave-particle duality of matter, that the probability wave of a quantum particle somehow allows it to appear on the other side of the barrier. A recent experiment with rubidium atoms further demonstrated this phenomenon, but offered no new insight into why it occurs.

Quantum entanglement and quantum tunneling become understandable if one remembers that space is quantized, not a continuum. Space is not infinitely divisible, its smallest dimension being the Planck length. Thus, returning to the analogy of the movie film, motion can be understood as a series of *states*, of *phases* of a quantum particle, just as each frame on a reel of film is a still photograph separated by a gap. Think of the celluloid of which the film is composed as the zero point energy. Even if there is no apparent motion between two frames of film, the film is still rolling, there is a connection between the succeeding frames, there is motion. Here the analogy breaks down a bit, but this is the point: even though a barrier blocks the quantum particle, it cannot block the zero point energy in which the quantum particle – a wave – is embedded, and the zero point energy carries both *energy* and *information* which allows the quantum particle to reconfigure on the other side of the barrier. The only difference between quantum entanglement and quantum tunneling is that in the former the barrier between two entangled particles is not a physical or energy barrier, but distance.

Thus, in a sense, motion can be understood as the transfer of information. Remember that a wave is by definition a periodic disturbance in a medium that carries energy from one point to another without permanently altering the medium. A quantum particle is a spherical standing wave in the medium of zero point energy. However, remember that zero point energy is a wave also that moves much faster than the objects embedded in it. If zero point energy carries information about a quantum particle from one side of a barrier to the other and allows this particle, in essence, to reconfigure itself, then motion in this instance is effected by a transfer of information.

Let us return to the analogy of the film. Let the length of film on the reel represent the
distance from here to a point in another galaxy. But let us say that there are only two frames on the film, one at the beginning, the other at the end, and the rest of the film is just one big shutter gap. But instead of running the projector and sitting through two hours of darkness before the fleeting initial frame reappeared, suppose we had a special film that, instead of being projected in the traditional way, could be scanned and projected in a small fraction of a second. Thus, we would see one still image in less than the blink of an eye *and* still go through the entire reel of film. If we could do that, then we would have an insight into how

faster-than-light space travel will be effected by humans, possibly in the not too distant future.

We are still in the infancy of space travel. We know that space travel as currently imagined is woefully too slow to carry humans even to the nearest star system. Yet we press on. Something deep in our psyche tells us that, somehow, we are going to travel to the stars. But one thing is for sure: interstellar space travel will be nothing like currently imagined by either scientists or science fiction writers.

Imagine a machine that could simply *translate* the information that makes up the quantum particles of our macro bodies into the zero point field and project it along a certain vector, say, to another star system, then translate the information back into its original form composed of atoms and molecules. Since zero point energy flows at twenty billion times the speed of light or more, we would essentially be hitching a ride, and would not even need a propulsion system.

Another possibility would be projecting a hologram, a simulacrum of ourselves through the zero point field to a distant location and communicate with it at zero point speed. In other words, we would be using something very much like the energy bridge between two entangled quantum particles.

Consider what manner of worlds we would discover were we to have the capability of zero point speed travel. What if we found a world whose conditions were suitable in every way for human colonization, whose temperature, atmosphere, gravity, chemical makeup, flora, fauna, and microbial life were so similar to Earth's that humans could live there with no problem. Assuming that another highly advanced human or humanoid species, or any intelligent species whatsoever, did not already live there, then travel there for humans would be acceptable. But assume that this same sort of planet was already inhabited by a primitive humanoid species, perhaps like us genetically but not as evolved as us. What then? Would it be acceptable to colonize such a planet? If we did, would it be wise to share our advanced technology with a species that had not yet achieved a level of civilization which would enable them even to comprehend such technology, let alone utilize it? Further, what would happen if we discovered planets with civilizations that were as advanced or more advanced than us? What then? Would we be hailed as fellow sapient beings or attacked as potential interlopers?

There are many possibilities. What if we encountered a world whose technology and culture were, say, on a par with early twentieth century Earth's, and the nations there were embroiled in the same types of wars and foibles that our ancestors went through? Do we intervene? Do we emerge out of the sky as saviors, or do we leave them be to unfold as we unfolded?

What if we found a young planet similar to the early Earth where life in any form did yet exist? Do we then introduce DNA in some manner and allow it to unfold into an Earth-like biosphere?

Now consider all of these scenarios. Did any of them happen here? Can it be that humans are creatures whose very nature it is to, at some point in their unfolding, contrive a way to travel to other planets? What are we humans, anyway?

If we can answer that question, then perhaps we will solve the riddle of the universe.

13

Has human evolution reached its apogee, or are we just some sort of transitional phase? Friedrich Nietzsche taught that mankind is not only a transitional species, but that our evolution into something higher, the *ubermensch,* would be a self-directed process. Consider that: *we humans, utilizing science, can deliberately change our very physical nature to conform to a predetermined purpose.* The related fields of transhumanism and posthumanism explore the possibilities and the ethics of such endeavor.

The term transhumanism was coined in 1940 by the Canadian philosopher W.D. Lighthall. The idea was expounded upon by the British biologist Julian Huxley in a 1957 essay, and Huxley is considered the founder of the movement. His basic premise is that, through science and technology, mankind could "transcend" itself. Huxley's main concern was the amelioration of the harshness and brutishness of human existence. But later thinkers saw this transcendence as leading to changes of such magnitude that they would give rise to beings of such a nature and capabilities that they could not properly described as human. Genetic engineering, digital technology, and bioengineering could create such modifications in the human species that it would indeed be a new species altogether. Possible scenarios include a blend of human and artificial intelligences – cyborgs – or completely synthetic artificial intelligences into which human consciousness could be "uploaded, " thus making an individual essentially immortal.

Immortality. Perhaps this is the true driving force behind

transhumanist and posthumanist thinkers. The curse of sentient beings is that they are aware of their own mortality. Humans have dreamed of immortality since before the beginning of civilization. Something deep in the human psyche yearns for a life beyond – and better – than this one.

Another idea to emerge out of posthuman theory is the notion of a "posthuman god," which asserts that at some time in the future humans will, through self-directed modification, become so advanced that they will be god-like by current human standards. Consider how this notion fits in with the idea of space travel at zero point speed. Posthumans arriving on a planet where humanoid creatures were still in a primitive stage would seem as gods. Could not the same thing have happened here? In this case, we would just be completing the cycle.

Arthur C. Clarke once observed that "any sufficiently advanced technology would be indistinguishable from magic." Just as humans have always yearned for immortality and mused about the gods, we have always dreamed of magical powers. Perhaps the fascination with magic is some deep atavistic memory of the seemingly god-like powers of the beings who directed our early evolution, or a yearning hidden deep in our DNA which compels us to achieve transhuman powers.

If all of this seems farfetched, remember that this planet will be habitable to human life for about another 900 million years. Look at how far we have come in just the 5,500 years since the beginning of civilization. Think how much farther we could advance in just another century, let alone a thousand years or so. And remember how limited our perspective is. In *Paradise Lost,* Adam relates to the angel Raphael how he awoke within the Garden of Eden having no idea who he was, where he was, or why he was there. Adam then goes on to ask many question of Raphael about God and the cosmos. Raphael admonishes Adam that he need not know these things, that some knowledge is for God alone. Like Adam, we still wonder about our origins and our purpose.

One clue to our purpose as a species is that we create

artificial intelligence. We have grown so adept at creating ever more powerful computers that many thinkers have expressed concern that artificial intelligence may supplant and dominate human intelligence. The development of a super powerful computer that could design ever more powerful computers could result in what transhumanists call a "technological singularity." Supposing that humans developed computers that calculated at zero point speed, it is very conceivable that the technological singularity would not be far off. Further, consider the ramifications of humans uploading their consciousness into the zero point field. Essentially we would have the capability of calculating at twenty billion times the speed of light and traveling at the same speed. We would truly be transhuman, would be like the god-like creatures who initiated and guided human intelligence in the first place. In other words, *we would have come full circle.*

But all of this begs the question: what is man? *Why* is man? Are we something that exists for its own sake, some art form, some curiosity, some plaything? Are humans some sort of vessels in which zero-point beings deposit themselves on their journeys through various iterations? Like Adam in *Paradise Lost* we find ourselves puzzled by our very existence, and
perhaps a bit vexed that the grand cosmic intelligence which created us leaves it us to solve the very riddle of our existence. In order to get a better understanding of what we are, let us consider the possibilities of the being or beings which engineered us in the first place.

14

Put aside all your religious and philosophical presuppositions, and imagine that you, like Adam, just awoke one day on this planet. Only you do not have God or the angel Raphael to give you instruction and answer your questions. However, unlike Adam, you know physics and astronomy and the mathematics to go along with it. You know about zero point, biology, chemistry, and cosmology. You accept the strong anthropic principle, and believe that life as we know it was engineered. So you are left with the question: what was the nature of the intelligence that designed the atom-based portion of the universe as well as carbon-based life?

Having studied Halton Arp's theories, you know that we are only observing a relatively small portion of the infinite universe. Once Arp discovered that the redshift did not indicate recessional velocity but newly created matter, he went on to deduce that the only portion of the cosmos we can be certain that we are observing are the Virgo and Fornax galactic superclusters, which lie within a 65 billion light year radius of earth. Thus, we cannot be sure that there are stars and galaxies beyond this radius. What we call the universe, the portion of space occupied by galaxies, nebulae, stars, and planets, may just be a tiny dot within the infinitude of space, of

the zero point field. The universe as we know it, and the carbon-based life forms within it, may be a total anomaly, a weird aberration. Far from being the center – and the purpose – of the universe, humanity may just be one odd attempt by a superintelligence to create what to it would be artificial intelligence. Thus, in our quest to understand our origins, we must first disabuse ourselves of any notion that the universe as we know it represents the universe as a whole, and that we humans occupy some special place within that greater universe.

Our starting point might be to consider the question: what can we infer about the force which engineered us just from the very fact that we are able to consider the possibility in the first place? First of all, this force would be intelligent and powerful enough to create matter from zero point energy in the cores of active galactic nuclei which would then explode, creating quasars which in turn would evolve into galaxies. From this newly created matter – hydrogen gas – stars would form, and over their life spans heavy elements would synthesize in their cores. Some of these stars would explode as supernovae, spewing these heavy elements into space which would then accrete around other stars, gradually forming planets, a few of which would be hospitable to human life.

As we discussed earlier, the laws of emergence suggest that such an intelligence could have arisen in the dense plenum of the zero point field. Are there many, perhaps an infinite number, of such intelligences? For example, is each galaxy a product of a separate super-intelligent being? Or is perhaps even each planet which hosts the unfolding of an intelligent species presided over by some zero-point based being, something prefigured in Hegel's idea of the *Weltgeist?*

Consider this hypothesis: under the right conditions a configuration arises in the zero point field that can perpetuate its own existence, in the process developing what we know as intelligence. Further, this configuration can channel and manipulate zero point energy in such a way as to create what we call matter, and, further, can use this matter to construct thinking machines – we humans. There is no reason to believe that there is any limitation on the number of such configurations that could exist within the infinitude of the zero point field. There is no reason to believe that such configurations do not interact, cooperate, and share information. Further, there is no reason to believe that we humans do not share somehow in the processing – and origination – of knowledge within this network of superintelligences.

The common thread is intelligence. Intelligence begetting intelligence, and so on *ad infinitum.* The human condition incessantly spurs us on to a greater understanding of the forces of nature in order that we may better harness these forces. Consider: if the universe is teeming with myriads of different intelligent life forms, all of them considering the universe from a unique perspective, then the universe is a place of eternal novelty and wonder. Those philosophers who, like Sartre and Camus, view human life as essentially meaningless, might do well to ponder the nature and the origin of that very phenomenon of intelligence which enables them to seek the meaning of life – or lack thereof – in the first place.

Our meaning lies in intelligence. Intelligence is our defining attribute. It is also the defining attribute of the force which created us, and of the "artificial intelligence" which we strive to create. Intelligence is a phenomenon in and of itself, something that exists for its own sake. Any intelligent life form has meaning just by virtue of its intelligence. The existentialist philosophers saw our intelligence as the very reason of the meaninglessness, the absurdity of our existence. In their view, life is just a funny accident, and we live and die on an insignificant planet orbiting a commonplace star in an infinite universe unaware and unconcerned with our existence. Our plight is that we, as intelligent creatures, are aware of our hopelessness. The flaw in their reasoning is that they assume that the force of intelligence is just an isolated given, that it somehow just evolved. However, if human intelligence is just one aspect, one link in the chain of an ever expanding sphere of intelligence, then the existentialist view needs serious reconsideration.

If we assume that the force of intelligence pervades the zero point field, and that humanity is an aspect of that force, then we have a better grasp of our very meaning. Further, when we consider how limited our perspective is, that we are but a tiny blip in the unfolding process of this cosmic intelligence, then the blanket assertions of the meaningless of life by the existentialists seem utterly presumptuous and baseless. In short, things may not be as they seem, human life may not be what it seems.

Our distant ancestors felt instinctively that there were life forms superior to humans. They were *animists,* they believed spirit beings existed alongside humans and other animals, often interfering in their affairs. Over the ages, animism gave rise to polytheism, the belief in many gods. The ancients' conception of gods was, in many cases, far from reverential. Gods were viewed as flawed creatures, as much motivated by such foibles as greed, lust, anger, and jealousy as humans. In some cultures, one god out of the pantheon was worshipped exclusively while the existence of other gods was acknowledged. This practice is called henotheism. For example, the proto Hebrews worshipped the Mesopotamian war god Yahweh. Over the centuries, Yahweh became the sole god worshipped by the Hebrews, but as late as the sixth century B.C. Jewish theologians acknowledged, albeit derisively, the existence of "foreign" gods which they labeled collectively as Baals.

The idea of monotheism began more as a philosophical concept than a religious one. It grew out of the principle of monism, proposed by Thales and Anaxamander, that behind the multifarious phenomena of the universe was a single basic substance or principle. The first recorded references to a unitary, monotheist God are from the works of Plato and Aristotle. It is no accident that later Jewish, Christian, and Muslim theologians were heavily influenced by Greek philosophy, particularly Aristotle. The medieval Jewish philosopher Maimonides defined God as a unity "unlike any other possible unity."

As we shall see, the idea of God as a unity would strongly influence Spinoza. Conceiving of God in this way is not so much monotheistic as pantheistic. God encompasses everything. Further, as we shall also see, this idea has some support in zero point theory as well as emergence theory.

Let us consider the possibilities.

15

Imagine the precarious situation Baruch Spinoza found himself in as a free thinker. A descendant of Portuguese Jews who had fled the Inquisition, he was born and raised in the tightly-knit Jewish community of Amsterdam, Holland. The Dutch were a tolerant people, but in the seventeenth century blasphemy, or what could be interpreted as such, could get one imprisoned. The Jewish community, ever cautious about offending the Dutch Christians under whose tolerance and hospitality they had found security, stability, and prosperity, were doubly sensitive to heretical notions. Spinoza, the highly intelligent son of a prosperous merchant, showed early prowess as a scholar, absorbing the wisdom of the Torah and the Talmud from the elders of the synagogue. They had high hopes for young Baruch, perhaps even seeing him as rabbinic material. However, our young scholar had an independent streak that would not only get him excommunicated from the synagogue at the age of twenty-four, but almost killed by a knife-wielding assailant.

His primary offense was rejecting the literal truth of both Hebrew and Christian scripture. He asserted that the miracles portrayed in holy scripture are metaphorical and allegorical, designed to attract as broad a base of adherents as possible. However, he maintained that the perceptive student of scripture can look beyond the metaphors and allegories of scripture to the underlying wisdom. Taking this skepticism as a starting point, Spinoza sought the truth of ultimate reality by reading broadly and deeply. He was greatly impressed by the medieval Jewish philosophers Maimonides, Hasdai Crescas, and Moses of Cordova.

In particular, Spinoza was taken by Crescas's notion that the material universe constituted the body of God, and Moses of Cordoba's assertion that God and the universe are one. Later, he found similar ideas in the philosophy of Giardano Bruno, who, thirty-two years before Spinoza's birth, had been burnt at the stake in Rome for his heretical notions.

We also know that he read widely in the Greek philosophers, and would have certainly been acquainted with Anaximander's idea of the *apeiron,* the boundless, irreducible substance from which all other forms of matter derive. He was also strongly influenced by Descartes, who taught that there was a homogenous substance that forms the basis for matter. However, Descartes also believed that there was a duality between mind and matter, and that mind or consciousness was also composed of an underlying substance. As we shall see, Spinoza rejected this dualism.

Consider the times in which Spinoza lived. The century before, Nicalaus Copernicus had asserted that the Earth orbited the Sun, not vice versa. In 1600, Giardano Bruno had been burnt at the stake for teaching this heresy, and forty years later Galileo had been sentenced to house arrest for proclaiming his astronomical observations that proved Copernicus's theories. In 1666, Newton published *Principia Mathematica Philosophae Naturalis,* which for the first time expressed the immutable laws of motion and gravity in elegant mathematics. In Great
Britain, free thinkers like John Locke, Algernon Sidney, John Lilburne, John Milton, and David Hume were boldly publishing works advocating freedom of conscience, speech, and the press. Sidney would be beheaded, Lilburne would have an eye poked out, and Milton, old and blind, would be imprisoned in the Tower of London, saved from beheading by influential friends. In Holland, cut off from his own people by excommunication, and fearing the wrath of Dutch political and ecclesiastical officialdom, Spinoza lived modestly, making a living as a tutor and a lens grinder. He only published two works during his lifetime, but kept up a lively correspondence with other philosophers. He came to the

attention of such luminaries as Jan de Witt, the Frenchman Prince de Conde, and the polymath Gottfried Leibnitz. He was even offered the chair of philosophy at the University of Heidelberg, which offer he graciously refused, explaining that his views might bring the university into conflict with the authorities.

Like generations of wisdom seekers afterwards, these luminaries were entranced by the originality and intricate development of Spinoza's ideas. Basically, Spinoza totally redefined the concept of God. The term pantheism was coined by a later thinker to describe Spinoza's metaphysics. Indeed, this in an apt term, but it requires clarification.

Spinoza based his metaphysics on the premise that there was only one basic, irreducible, indivisible, infinite, and eternal *substance* in the universe. The word "substance" derives from the Latin – in which language Spinoza wrote – and literally means "that which stands under." It did not have the materialist connotation we commonly ascribe to it. In meticulous geometric logic replete with axioms, propositions, and proofs, Spinoza asserts that underlying all the ephemeral forms or "modes" we observe in the universe, there must be some uncaused principle of which these are just phenomenological manifestations. He called this substance *Deus sive*

Natura – God or nature. He reasoned that this substance has infinite "attributes", only two of which could be grasped by the human intellect: thought, i.e. intelligence, and extension, i.e. physicality. In positing the limited nature of human perspective, Spinoza actually broadened human perspective. Whether or not one accepts Spinoza's metaphysics, one has to accept his admonition that our ideas about God, the universe, and humanity may be flawed, and we have to take pains to look beyond them. In doing so, Spinoza became one of the most influential philosophers in history, and, in some quarters, the most hated.

Spinoza's God is simply the totality of all that is. God, or nature, is a unity, a holistic being that unfolds according to its intrinsic essence. Human intellect is but a part of this whole, but Spinoza points out that God is not concerned with humanity and not consciously involved in its affairs. God is impersonal, and much of our misunderstanding of him arises from ascribing human qualities to him. God is a totally rational being, and, for Spinoza, morality consists of mastering ones passions, seeing beyond the fleeting ephemerae of human life, and living in accord with that same reason which animates God and nature. Living rationally leads to true freedom. Spinoza's political ideal was democracy, for in a free democracy rational beings could work, trade, interact peacefully, and thrive.

We find strong echoes of Spinoza in German idealist philosophy, particularly Hegel; Romantic Age literature; Deism, a philosophy which strongly influenced Washington, Jefferson, Franklin, and other Founding Fathers; and Transcendentalism, an American intellectual movement of the nineteenth century. The appeal of Spinoza owes much to its logical structure. If one accepts his premise that there is one and only one irreducible substance in the universe, then his conclusions follow naturally. Such a view has strong appeal to a rational mind seeking an answer to the riddle of existence. But one wonders: if Spinoza were alive today, what modifications would he make to his philosophy in the light of our present scientific understanding?

Perhaps philosophy itself needs a rebirth. From the birth of Western philosophy in Greek Ionia in the sixth century B.C, until the breakthroughs in empirical science in the seventeenth, eighteenth, and nineteenth centuries, philosophy, especially metaphysics, relied on deductive reasoning from nebulous premises. Philosophy was like unto someone groping in the dark in an unfamiliar setting, trying to imagine scenarios that would explain his predicament. But with the advent of Copernicus, Galileo, Newton, Faraday, Maxwell, and Darwin, the dark gave way to light. More and more, thinkers came to eschew speculative metaphysics for empirical science. Even those thinkers who discerned the guiding hand of some superior intelligence behind nature and mankind resigned themselves to the fact that a being was beyond the ken of human understanding, and regarded speculation on the topic as pointless. But is it?

Were Spinoza alive today he might very well find in the zero point field the irreducible substance *Deus sive Natura.* He might find in the anthropic principle mathematical confirmation of his belief in an intelligence greater than human intelligence. Further, he might make certain adjustments to his original overall conception of God. Oddly, in Spinoza's system, humankind is not so much a conscious creation of God as a mode of two of God's attributes: thought and extension. Humankind exists because it must necessarily exist according to the very nature of God. Further, and more incredible, Spinoza conceived of God as not only impersonal, but not even consciously personal as we understand that quality. God is what he is and brings about what he brings about with the unconscious impersonality of a force of nature.

In contemplating Spinoza's approach to God, one must remember that Spinoza was battling the ingrained anthropomorphic notions of God that were being upheld by a powerful, often oppressive religious establishment. His was no easy task. Not only was he arraying himself against powerful institutional forces, he was trying to overcome the intrinsic limitations of human perspective itself.

What is remarkable about Spinoza's philosophy is that, aside from a few modifications, it could very well serve as a framework for a workable, credible metaphysic that is not only in accord with empirical science, but offers a logical explanation for all those things that, so far, science cannot answer. In short, he was on the right track. But one may ask: what difference does any of this make? We can speculate on these matters forever and still not prove them empirically. Perhaps so. But just perhaps, in the not too distant future, when we tap into the zero point field, we may also be tapping into the infinite intelligence that created us, of which we are a part. Remember: here on this small planet we are only perceiving an infinitesimal fraction of the whole.

Let us consider the possibilities.

16

Imagine that we have found a way to tap into zero point energy and can travel, communicate and compute at twenty billion times the speed of light. The first earth-like planet we travel to is still in the stone age. Oddly enough, the inhabitants there are not only human-like, they are genetically as human as we are, indeed are the same species as we Earth humans. That brings us to our first conundrum. How is it possible for a distant planet to have beings that are spun from the same DNA as us? Our first conclusion would be that humanity, or DNA-based life in general, is not limited to Earth. So we ask ourselves a series of questions: was DNA planted on Earth by an alien visitor long before, or is it the means whereby some higher advanced intelligence brings about intelligent beings on countless suitable planets throughout the cosmos? What is the nature of this higher intelligence?

While we are pondering this riddle, we interact with the primitive souls living on this planet. Their features are a bit coarser than ours, but they speak a language that, our computers tell us, has phonetical similarities to Proto-Indo-European, the mother language of all the European languages. Further, the grammatical structure of their language is quite similar to Proto-Indo-European. But how can this be? Brain scans of some of these people reveal that the language center of their brains is as developed as ours. The linguists and the physiologists on our expedition see this as confirmation of something they have suspected all along: humans, living in society with other humans, will develop a language according to the genetic guidelines within the language center of the human brain. *This is how we are engineered.*

As we interact more and more with these primitives, we notice that although they hold us in awe and are quite fascinated by us, they are not overly stunned by our arrival. Through our computer translation software, we learn of stories passed down from their ancestors telling of other visitors from afar. We also learn that, although these people are loosely organized into separate clans and tribes, they all live in proximity to a great inland sea in a temperate zone of this planet. They hunt, gather, and fish, but have not yet discovered agriculture or animal husbandry.

After a lively debate among members of our expedition, it is decided that perhaps it would not hurt to give these fellow humans a gentle nudge toward their next stage of development. In the grasslands around the inland sea great herds of horses flourish. Painstakingly, working through pictograms and the translation software, one member of the expedition who is an experienced horseman explains to the young bucks of the people how to break and train horses, not just for riding but for pulling wagons. Then he remembers: now he has to teach them what a wheel is. The members of the expedition are stunned by how quickly the people learn how tame horses, to ride, and to fashion wood into wheels and wagons.

They explain to the primitives: "Just as it happened on our world, so it will happen on yours. You will spread westward, to the great fertile plains and river valleys. Your tribes will each go their own way and settle new lands, and their progeny will wander and settle even further. Then, a long time from now, your seed will spread to other worlds, just as we have done. We are of the same blood and flesh. It is the way of us, of us humans."

After bidding farewell to our primitive friends, our expedition moves on to a planet that has been all but destroyed by a thermonuclear war. They do not tarry long, judging that this particular unfolding of humanity was a failure. The next stop is a planet whose civilization is actually a bit more advanced than ours. Everyone is happy, healthy, and at peace with everyone else. Our expedition is welcomed warmly by the scientific establishment there. They explain that they have been expecting us. They wear translation apparatuses that convert their speech into English. One man in particular, a very striking, imposing man, explains that on many occasions he had been part of expeditions just like ours.

"I was at Earth over six thousand years ago. I taught your Proto-Indo-Europeans how to tame horses. And now here you are!"

"But how can that be?" they ask.

He explains: "I am a zero point being. Although I did not realize it at the time, I was in human form for a purpose. And now all of you have traveled through the zero point field. You have morphed into zero point energy and back again. That is just the first part of the journey. And now that you are here, you will become like me, like all of us . . . ."

Does all of this sound farfetched? Once we consider the properties of the zero point field it actually seems quite plausible. This scenario explains what we are, why we are, how we originated, and what our future holds. It is a synthesis of empirical science and a logical hypothesis derived therefrom. Philosophy has come full circle.

Consider the implications of this scenario in the light of Spinoza's metaphysics. In his view the totality of the zero point field and all the modes that exist within it would constitute a holistic being that is more than the sum of its parts. Intelligence within intelligence within intelligence yielding an infinity of new emergent properties, new stages of being. But yet one wonders: is there something beyond, or more accurately, prior to the zero point? Is God an emergent property of the zero point field, or is the zero point field an artifice of some force, some intelligence that is beyond the powers of human imagination?

Plato could not quite conceive of a God who deals directly with, let alone be a part of, the material universe. He invented the concept of the Demiurge, literally "divine craftsman," a being that mediates God's will in the material universe. Is there something more basic than the zero point field, something prior to it? Perhaps once we have tapped into the zero point field and partake of cosmic intelligence, then we can answer that question. But the fact that we can even conceive of the question reveals much about our nature.

For now, we must concentrate on learning how to tap into and utilize zero point energy. The first hurdle in achieving that goal is getting able physicists to investigate this field. Many orthodox physicists scoff at the idea that zero point energy can be utilized. Many, if not most, physicists still cling to the Big Bang theory, the quark theory, and the corollary of relativity that mothing can travel faster than light. However, facts are stubborn things, and if the model of the universe suggested in this volume is indeed in accord with objective reality, then the zero point world is not far away. We will know the truth because that is what we were designed to do.

If Spinoza were alive today he might revise his metaphysics somewhat. His new, revised *Deus sive natura* might be a being with a bit more awareness, a bit sharper intentionality than his original conception. Everything about our scale of the universe indicates a conscious, assiduous, creative, *involved* intelligence. Perhaps we humans should conceive of ourselves as not only an aspect, an atom of a holistic cosmic intelligence, but as individuals in a symbiotic with same. Spinoza's view that mankind exists because we are an inescapable byproduct of the universe, that things just could not have been otherwise, is an evasion, an expedient. The assertion that things are what they are because they could not have been otherwise is a tautology unworthy of Spinoza's intellect, and is perhaps the weakest link in his metaphysics. In his defense, he did not have the knowledge at his disposal that we have now.

It is also instructive to wonder what modifications Hegel would make to his metaphysics were he alive today. In spite of his deliberate obfuscations, Hegel saw the world as an unfolding process and tried to comprehend both the mechanism and the purpose of this unfolding. He envisioned a stage of human development in which humankind would have a full understanding of the unfolding of which they were a part. Hegel would recognize his *Weltgeist* – World Spirit – in the cosmic intelligence we have hypothesized in this volume.

And let us not forget old Aristotle. He would probably concede that his Unmoved Mover is not so static after all, but is a being whose very essence is movement, is energy, a being of infinite energy in infinite time and space. A man of empirical science and mathematics as well as metaphysics, Aristotle would be quite gratified that the sublime, infinite intelligence that he intuited behand the wonders of nature is discernible within the framework of science and mathematics.

But what of us, us everyday mortals? What are we to take away from all of this? First of all, that things are not what they seem, the universe is not what it seems. When we despair, or are perhaps just bored and frustrated with life, we need remind ourselves of the very improbability of our being here in the first place, of that great cosmic intelligence *of which we are a part.* When life seems meaningless, just consider: we live in a universe of infinite energy, and therefore infinite possibilities. At such times, simply consider the possibilities.

APPENDIX

The Science

In the foregoing we have presented just enough science to explain our hypothesis and put it in context. In the following, we will give a much more detailed exposition of the science upon which our hypothesis is based. In particular, we will discuss zero point theory, Planck units, Tom van Flandern's discovery of the speed of gravity, Halton Arp's continuous creation cosmology, Milo Wolff's theory of subatomic particles as spherical standing waves, the science behind the strong anthropic principle, the biological evidence for intelligent design, and emergence theory.

Zero Point Energy

It was once believed that at a temperature of absolute zero – zero degrees Kelvin – all atomic and molecular motion ceases. However, when it was discovered that this is not the case, German physicists investigating the matter coined the term *nullpunksenergie* – zero point energy. By definition zero point energy is the lowest possible energy state that an atom or a molecule can
have. Even at absolute zero atoms and molecules have vibrational motion. The concept of zero point energy was developed by Max Planck as he expounded on his discovery in 1900 that light is not emitted in a continuous ray but in discrete units called "quanta." The idea of zero point energy soon became associated with the idea of an aether, a medium pervading space that allowed not only the propagation of light, but, as Einstein was to assert in 1921, gravity as well.

Zero point theory gained traction when it was demonstrated that it explained certain hitherto unexplained anomalies. For example, at absolute zero helium remains a liquid. According to standard theory it should become solid. Further, according to standard theory, an electron orbiting an atomic nucleus should lose energy and fall into the nucleus. However, the fact that it did not indicated that the electron was drawing energy from some source.

Spontaneous emission was another troubling phenomenon explained by zero point theory. By definition, spontaneous emission is the process by which a quantum particle emits a photon as it changes from a higher energy state to a lower one. What caused this emission? In 1927 Paul Dirac published his theory of quantum emission and absorption in which he made the astounding assertion that the "vacuum" is a plenum of energy. In his theory, when a photon is emitted it "jumps" from the "zero state," and when it is absorbed it "jumps" into the "zero state." He further explained that "since there is no limit to the number of light quanta that may be created in this way, we must suppose that there are an infinite number of light quanta in the zero state." In 1935 Victor Weisskopf wrote that it logically follows from quantum theory that there are "zero point oscillations of empty space."

Later, Hendrik Casimir discovered the eponymous effect whereby two electrically neutral plates placed closely parallel to one another would experience an attractive force between them caused by the fact that there was more vacuum fluctuation energy outside the plates than between them.

As time went by, other physical anomalies were discovered which were neatly explained by zero point theory. The Lamb shift, in which the two energy levels of a hydrogen atom have different energy levels when both should have the same, is explained by an interaction with the zero pout field.

Another puzzling fact cleared up by zero point theory was the anomalous magnetic moment. Being a charged particle, an electron creates a magnetic field when it rotates and when it orbits an atomic nucleus. This phenomenon is called magnetic moment. However, the observed properties of the electron's magnetic moment did not jibe with Dirac's calculations. There was something interacting with the electron – the zero point field.

Another anomaly discovered in 1975 was Delbruck scattering, in which photons in the Coulomb field of an atomic nuclei were scattered by *vacuum polarization,* which is caused by the spontaneous appearance of electron-positron pairs out of the vacuum that changes the original distribution of charges and currents that generated the original electromagnetic field. One has to assume the existence of the zero point field to explain vacuum polarization.

Vacuum birefringence is another anomaly explained by zero point theory. Birefringence is a property of certain crystals which allows them to split a beam of light into two different beams which travel at two different speeds in two different directions. However, under certain conditions, birefringence can occur in the vacuum itself. When a portion of space contains a magnetic field above a certain critical point, the perturbed area becomes birefringent, such that light waves traveling through it not only become split into different directions, but travel at different speeds, *some of them significantly less than c, the assumed universal speed of light.* Because of the great amount of energy it requires to make a portion of the vacuum birefringent, this phenomenon is observed by studying the light traveling from or passing by massive stars.

So what is the nature of this phenomenon known variously as zero point energy or vacuum energy? What are its properties? Why have physicists not done more to describe it mathematically? Indeed, some physicists seem to avoid the topic altogether because assuming the existence of the zero point field does not jibe with certain orthodox notions of standard model physics. The principles of quantum electrodynamics suggest that zero point energy is very dense – 10^{113} joules per cubic meter, more energy by far that is released by all the electrical plants and nuclear reactors on Earth. But what is its wavelength, its speed of propagation?

The key to answering these questions is the Planck units, natural units of length, mass, time, and temperature derived from the fundamental constants of nature h – the Planck constant, which is the smallest known unit of energy, g – the gravitational constant, which gives the gravitational force between two bodies, c – the speed of light, and k – Boltzman's constant, which relates the absolute temperature of a gas to its pressure and volume.

Planck Units

In order to grasp the concept of the Planck units, let us first discuss the Planck constant h. Max Planck derived this value – 6.626 x 10^{-34} joules per Hertz – from empirical observations he made of light emitted from a black body, in this case the interior of a black metal box with an aperture on one side. Although a perfect black body absorbs the light which hits it, it emits light when heated to certain temperatures, hence the terms "red hot" and "white hot." Using a spectrograph to measure the frequency of the light emitted by the black body at different temperatures, one would have expected to see a steady increase in frequency directly proportional to the increase in temperature of the black body. Graphed, this relationship should have formed a single continuously rising line or curve. Instead, the graph formed a series of *ever steeper bell curves one atop the other demonstrating that light was not emitted in a continuous wave* but in discrete packets or *quanta.* How could this be?

Planck explained that light is produced by "harmonic oscillators," which are electrons, atoms, and molecules which vibrate at certain frequencies measureable in multiples of h such that the energy E of a harmonic oscillator equals h x f, where f is the frequency of the oscillator. The discovery that light is quantized and measureable in terms of h revolutionized physics. In 1905 Einstein used the new theory to explain how light quanta – photons – stimulate the flow of electrons in certain materials instantaneously upon striking the material. If light were a wave, he reasoned, the flow, known as the "photoelectric effect," would not be instantaneous. However, in some situations, light does indeed behave as a wave. By way of illustration, Einstein likened a light wave to a line of soldiers and a photon to an individual soldier. The line can compress or elongate, can be thrown off course by collision with another line, but the soldiers remain individual entities.

Thus, by empirical observation, Planck derived a new constant of nature h to go along with those already discovered – g, c, and k. Pondering this fact, Planck realized that he could arrange these constants mathematically in ways that would yield natural units of length, mass, time, and temperature that could supplant the old arbitrary units such as feet and meters, pounds
and kilograms, and degrees Fahrenheit, Celsius, and Kelvin. The derivations were made thus (note: the symbol h with the crossbar is known as the "rationalized Planck constant;" it is simply h divided by 2 pi, which is one radian, and thus the rationalized Planck constant is useful in describing periodic motion such as sine waves and rotation.):

$\sqrt{\dfrac{\hbar c}{G}}$ 1.62×10^{-35}m -- the Planck length

$\sqrt{\dfrac{\hbar c}{G}}$ 2.18×10^{-8}kg – the Planck mass

$\sqrt{\dfrac{G\hbar}{c^3}}$ 5.39×10^{-44}s -- the Planck time

1.41×10^{32}K -- the Planck temperature, which is derived by multiplying the Planck mass by c^2 -- Einstein's mass/energy conversion, and dividing the product by Boltzman's gas constant.

Contemplating these values, many physicists have wondered if they have significance beyond being mere "natural units." As we have seen, the Planck length makes all the sense in the world if one assumes that it is the minimum wavelength of zero point energy. The Planck time, which is the time it would take for light to traverse the Planck length, nicely explains the fact noted by Einstein that time itself is quantized, thus allowing for the phenomenon of time dilation. The Planck mass is a bit anomalous in relation to the other units. While the Planck length and Planck time are by definition the smallest units of length and time imaginable, the Planck mass is

by comparison huge, being about the mass of a gnat. However, this quantity gives a strong clue to the nature of gravity. If gravity is an emergent property whereby zero point energy is diverted by mass, then the Planck mass can be understood as the critical point at which this diversion occurs.

Less clear is the significance of the Planck temperature. To be sure, it indicates that a great deal of energy resides within the fabric of space. It has been suggested that this temperature is some sort of critical point, for example what might exist within the core of an active galactic nuclei before it explodes, ejecting new matter that forms into a quasar.

The salient point is this: if one postulates the existence of the zero point, the significance of the Planck units becomes clear. As we shall see, this same postulate explains other vexing phenomena as well.

The Speed of Gravity

A corollary of Einstein's theory of relativity is that nothing can travel faster than light. However, as any student of celestial mechanics can attest, when computing the gravitational attraction between celestial bodies, be they stars and the planets orbiting them, or stars orbiting a galactic core, or distant binary pulsars orbiting one another, one has to treat the speed of gravity as instantaneous. As the astrophysicist Tom van Flandern discovered as a graduate student at Yale in the 1960's, physics professors and theoretical physicists alike tended to sidestep this issue. For example, we know that the light from the sun takes 8.3 seconds to reach Earth, and thus the apparent position of the sun we see is not its actual location, but 8.3 minutes ago. However, we also know that the gravitational force of the sun is coming from its true position. There is no discernible time lag. Further, it can be demonstrated mathematically that if gravity traveled at the speed of light the time lag would make planetary orbits unstable. The planets would go flying off into space on a vector tangential to the orbit thus interrupted.

One might ask: why is this a problem? If the available facts , both empirical and mathematical, suggest that gravity propagates at a much higher speed, perhaps infinite, than gravity, why is it so hard for some physicists to accept? Einstein himself sidestepped the issue by asserting that gravity is not a force at all, but "the geometry of space-time." In other words, massive objects curved the space in which they were embedded, and objects traveling near them followed this curve. However, in spite of this assertion, there are physicists who will twist themselves into knots trying to prove that gravity propagates at the speed of light. When Tom van Flandern's observations led him to the conclusion that gravity propagates *at least* 20 billion times the speed of light, he was called a "crackpot" and other unflattering epithets. When one's entire career and scientific reputation is invested in standard theory, an integral part of which is relativity, then one may become emotional and irrational at having that theory challenged. More on that later.

So how did van Flandern go about determining the speed of gravity? In one experiment he used precise timing data from stable pulsars in the sky to measure the acceleration vector for the Earth's motion around the sun's orbit, and then determined whether or not this vector was parallel to the direction of the photons arriving from the sun. The experiment demonstrated that they were not parallel, and, thus, that the light from the sun was traveling much slower than the gravitational force.

Another observation that bothered van Flanden was the fact that binary black holes were observed to exert gravitational force upon one another. However, according to standard theory, black holes are so massive than nothing can escape their event horizon because nothing can travel faster than light. But if gravity travels at the speed of light, how can it escape the black hole and affect another black hole?

Studying the gravitational effects that the two stars in the binary pulsar system PSR1534+12 had on one another, van Flandern arrived at the conclusion that gravity updates *at least* twenty billion times the speed of light. He published these results in the prestigious peer reviewed journal Physics Letters A after receiving assurances from an editor there that the paper would not be rejected because it contradicted conventional physics doctrine.

Consider: orthodox physicists who reject van Flandern's conclusions because they contradict Einstein's dictum that nothing can travel faster than light do so without even defining what gravity is. First of all, unlike the electromagnetic force which both attracts and repels, gravity only attracts. Further, it is many order of magnitudes weaker than the electromagnetic force, but operates over much, much greater distances. It is manifestly of a different nature from electromagnetism, which cannot exceed the speed of light. This essential difference is easily understood if one assumes that while light and other electromagnetic phenomena are wave-like effects *in* the medium of the zero point field, gravity is the diversion *of the flow of* the zero point field. Thus, assuming the existence of the zero point field nicely explains the nature of gravity, and the observed speed of gravity tells us much about the speed of zero point energy.

The hallmark of a sound theory is that it accommodates hitherto anomalous empirical observations without requiring adjustments to the theory. Thus far in our investigation, zero point has held up in grand fashion. How does it stack up against standard cosmological theory – the Big Bang theory?

Continuous Creation

In recent years the Big Bang theory has lost many adherents due to some troubling empirical observations, particularly those

THE PHYSICS OF SPINOZA'S GOD

which indicate that the universe is much older than the 14 billion years suggested by the theory. But even before these inconvenient observations, the Big Bang theory was on shaky conceptual ground. Foremost was the idea that, at some point, the universe exploded out of nothing. As discussed earlier, this notion arose out of our limited perspective and certain epistemological preconceptions, particularly our belief that, somehow, everything that exists had to have had a specific origin in time.

The big bang theory gained traction when Edwin Hubble discovered that the light from distant galaxies was red-shifted. If the red shift was caused by the Doppler effect, which occurs when the source of light is moving away from the observer, thus increasing the wavelength of the emitted light, then one had to assume that the galaxies were flying apart from one another and the more distant the galaxy, the faster the recession. Many astrophysicists and astronomers pointed to Hubble's findings as proof of the Big Bang theory, and built their cosmological models on the assumption that Big Bang theory was established fact.

However, in the 1960's the astronomer Halton Arp, an assistant to Edwin Hubble, made some puzzling discoveries about the recently discovered celestial objects called quasars, an acronym for "quasi-stellar objects" They were very bright, but also very red-shifted, which led **astronomers, reasoning that high-red shift indicated that were very far away, that they were very massive objects. Basically, they were anomalies, and, according to standard astronomical theory, unexplained. The mystery was solved when Halton Arp made numerous, documented observations that not only were quasars not as distant as assumed, but were always related to companion galaxies of a much lower red shift. Further, it was obvious that the matter which made up the quasars was ejected from these companion, or, more accurately, parent galaxies.**

Thus, Arp concluded that redshift does not necessarily imply recessional velocity. But what else could cause it? Adducing the Hoyle-Narlikar theory that particle masses can vary in space and time, he suggested that quasars, being composed of new matter, did not contain as much energy as older matter, and thus had a higher redshift than more energy-dense matter.

Consider how Arp's theory is explained by and supports zero point theory. If matter is created from zero point energy in the cores of massive galaxies, and then ejected, absorbing more zero point energy over time, then Arp's theory – continuous creation theory – both supports and is explained by zero point theory.

Continuous creation theory gained further support, and Big Bang theory another blow, when Arp discovered that redshift was quantized, that is, it assumed specific values, not the continuum of values one would expect if it were caused by galaxies flying apart from one another as a result of a primordial explosive creation event. Arp arrived at this conclusion by studying the z numbers of light from quasars. Z numbers are derived by comparing the degree of redshift from the expected light spectrum of a galaxy. Arp discovered that light from quasars took on the specific z numbers .061, .30, .91, 1.41, etc.. If redshift were caused by recessional velocity, then quantized redshift meant that the Earth was at the center of concentric shells of galaxies all moving away from one another – obviously not the case. Arp realized that redshift quantization could be explained as a function of electron spin, and mass could be explained as a function of the frequency thereof according to the formula $m = h/c^2 v_c$ where v_c is the Compton frequency of the electron.

Arp's discoveries caused a great deal of angst in the astrophysics community. His observations overturned decades of accepted theory and the professional reputations based thereon. He found himself unable to get his papers published in peer reviewed journals and book telescope time at the major observatories at which he had long been a fixture. It is a tribute to the intelligence and professional integrity of the astrophysics community in Germany that Arp was invited to continue his investigations at the Max Planck Institute for Astrophysics in Munich.

Anything that tends to undermine the Big Bang theory discredits other constituent theories of the currently accepted standard model as well, particularly the quark theory. Kwan Chiang, a physicist at the Imperial College in London, explains that the standard model is "patched up from many models and theories." The quark theory asserts that hadrons, which are particles that are affected by the strong nuclear force that binds atomic nuclei together, are formed from smaller particles called quarks. Hadrons include protons and neutrons, and very short-lived particles called hyperons that are more massive than protons and exist only in the artificial conditions of the laboratory, as well as mesons, also short-lived particles that are less massive than protons. Supposedly, quarks have fractional electric charge, spin, and rate of decay. The quark theory is quite complicated, and quite arcane. In addition to quarks, the theory posits the existence of particles called gluons that mediate the strong nuclear force which holds the atomic nucleus together against the repulsive electrical force the positively charged protons exert on one another.

Further, in order to explain why free quarks have never been observed, quark theory contains a corollary called "quark confinement," which asserts that quarks can only exist bound to other quarks in hadrons. This phenomenon is supposedly occurs because the free quarks and gluons existed in a super-hot plasma just before the Big Bang, but bound together as the expanding plasma cooled and matter as we know it formed.

Kwan Chiang has developed a theory of spatial structure that makes the quark theory superfluous. In explaining his theory, one of the objections he raises to standard theory is that it is incredibly complex and needs very rigid parameters to work. One of these is that it requires "micro-dimensions" within space to work. His theory asserts that three-dimensional space suffices to explain hadrons. As we shall see, if one assumes the existence of the zero point field, matter can be explained very simply as spherical waves.

Space Resonance

Milo Wolff, building on the work of the nineteenth century mathematician and physicist William Kingdon Clifford, formulated a theory of matter based on the simple premise that so-called "particles" like electrons and protons are really spherical waves in the medium of space composed of an "in" wave and an "out" wave superimposed on one another, and the wave center, which marks the point where in waves are radiated back outward as out waves is what we conceptually label a particle. Wolff explains the strong nuclear force by the overlap of the dense inner waves of the protons and neutrons in the atomic nuclei.

Notice how the empirical observations of Planck, van Flandern, Arp. and Wolff fit
together perfectly like the pieces of a puzzle. The model of the universe suggested by these constituent theories is, especially in relation to Big Bang and quark theory, elegantly simple. Also unlike the Big Bang and quark theories, this model does not need periodic tweaking to fit new scientific discoveries. To sum up, this model, which we will call the zero point model, asserts the following:

1 – Space is a dense plenum of energy whose characteristic wavelength is 1.62×10^{35} meters – the Planck length. The speed of this energy is at least 20 billion times the speed of light.

2 – The subatomic particles that make up matter are spherical standing waves in this energy, this zero point field.

3 – The speed of light is a function of the uniform nature of the zero point field just as the speed of a falling row of dominoes a function of the uniform nature and spacing of the dominoes. In this analogy the Planck constant h is equivalent to the minimum force it would take to tip over the first domino.

4 – New matter is created in the dense cores of certain galaxies and spewed into space to form what we call quasars. The high red shift of light from quasars is a function of its newness,

the fact that its matter waves have not absorbed as much energy from space as older matter.

5 – Gravity is the diversion of zero point energy by matter. It is an emergent property that does not become apparent at masses less than 10^{-5} kilograms – the Plank mass. Gravity updates at the speed of zero point energy – at least twenty times the speed of light.

6 – The speed of zero point energy explains certain faster-than-light phenomenon such as quantum teleportation, and suggests the possibility of faster-than light computing, communication, and travel.

7 – Certain natural phenomena such as the Casimir effect suggest that zero point energy can be tapped for human utilization.

Consider: if the zero point model is correct, then we live in a universe that is, at its most basic level, not only quite simple, but quite comprehensible. When we consider the principle of emergence, we begin to understand how such simplicity can give rise to that ultimate complexity: the force of intelligence.

Let us consider the evidence for a cosmic intelligence that engineered not only life, but the physical constants that made such life possible.

Intelligent Design

The very phrase evokes eye-rolls and condescending sneers from many in the scientific establishment. In spite of all the evidence in biology and physics that life was engineered and the laws of nature were fine tuned to allow for the emergence of life, the idea of intelligent design is taboo to the scientific elite, is branded as pseudoscience and its adherents dismissed as crackpots. Perhaps this attitude would be understandable if it could be demonstrated that adherents of intelligent design were skewing their observations and calculations to fit the presupposed notion that nature shows signs of intelligent design, but this is not the case. To the contrary, especially in the field of physics, it is mainstream scientists who are expressing dismay at, and trying to explain away, the fact that the constants of physics are uncannily within exact ranges so as to allow for the emergence of intelligent life.

The notion that the universe, or, more accurately, that portion of the universe in which we
find ourselves, was engineered by some vast intelligence carries with it certain emotional and cultural baggage. Those who remember the persecution of Galileo and others who went against the religious establishment see any appeal to design as irrational and unscientific. They see it as an appeal to the realm of the supernatural, i.e., as something which is by definition outside the ken of science and, therefore, not a valid subject of scientific investigation. But is this a valid assumption? If the evidence should suggest that the universe and life were engineered by a conscious intelligence, does it logically follow that such an intelligence is outside the realm of human comprehension? Is the term supernatural even a valid term? Is there a realm of being that exists outside nature, outside the universe, and is thus impervious to human reason, or can it be that there is no such thing as the supernatural, *just that portion of nature which we do not yet understand*?

The biochemist Michael Behe, whose book *Darwin's Black Box* sparked much interest in, and reaction against, the idea of intelligent design, summed up his approach thus: "The conclusion of intelligent design flows naturally from the data itself, not from sacred books or sectarian beliefs. Inferring that biochemical systems were designed by an intelligent agent is a humdrum process that requires no new principles of logic or science. It comes simply from the hard work that biochemistry has done over the past forty years, combined with the consideration of the way in which we reach conclusions of design every day."

Indeed, during the 1950's, it was a biochemist named Stanley Miller who tried to demonstrate that life could have arisen spontaneously on the early earth by the formation of amino acids

from the chemical compounds in the hydrogen-based atmosphere of those days. He
introduced hydrogen, nitrogen, sulfur and other chemicals into a glass sphere, subjected the
mixture to electrical discharge to simulate the lightning strikes that, supposedly, catalyzed the formation of amino acids on the primordial Earth, and studied the results. Indeed, a few amino acids formed. But subsequent, more elaborate experiments failed to reproduce any proteins, which are chains of amino acids. The proteins which make up living things are quite complex. Not only are they composed of over thirty different types of amino acids, they have to be in exact sequences of same. Further, proteins are often folded into complex shapes and have chirality, i.e., they come in right-handed and left handed versions. In the seventy years since Miller's classic experiment, scientists have yet to demonstrate how even the simplest protein, let alone the simplest strand of DNA, could have appeared spontaneously on the early Earth.

In *Darwin's Black Box,* Behe discussed the principle of "irreducible complexity," which states that complex biological structures like the eye, which are composed of distinct different parts, could not have evolved by chance because all their constituent parts, which make sense only if part of an integrated whole, would had to have had separate origins. For example, if just one constituent part of the eye, like the lens, corneal muscles, vitreous fluid, or retina is missing, the eye will not function. Darwin tried to explain the complexity of the human eye by positing that it evolved from the eyes of simpler life forms, the simplest of which were merely light sensitive neurons on the nerve ends of marine creatures. However, Behe demolished this argument by explaining how incredibly complex the process of sight is on the biochemical level: "When light first strikes the retina, a photon interacts with a molecule 11-cis-retinal, which rearranges within picoseconds to trans-retinal. The change in the shape of retinal forces a change in the shape of the protein, rhodopsin, to which the retinal is tightly bound. The protein's metamorphosis alters its behavior, making it stick to another protein called transducin. . . "

Behe goes on to describe several more links in the process until, finally, an imbalance of sodium ions across a cell membrane causes a current to be transmitted down the optic nerve to the brain. Consider how complex this process is, how complex are the molecules and structure which make it possible, and how if just one link in the process were missing or did not function, the brain, itself an incredibly complex structure, could not interpret the impulse as a visual image.

Behe describes other such irreducibly complex phenomena as well. For example, blood clotting. Behe describes a process that requires at least fifteen chemical reactions to cause blood to clot on a simple finger cut. Even more complex is the structure and function of cilia, hair-like structures that move fluid over a cell's surface or propel single cells through a liquid. The structure of a single cilium is thus: a bundle of fibers called an axoneme; each axoneme contains a ring of nine double microtubules surrounding two central microtubules; each outer doublet consists of a ring of thirteen filaments fused to an assembly of ten filaments; the filaments of the microtubules are composed of two proteins called alpha and beta tubulin; the eleven microtubules forming the axonemes are held together by three types of connectors . . . and so on and so on, such that the cilium functions by means of chemical power that allows a structure called a dynein arm on one microtubule to interact with another dynein arm on another microtubule. Again, note how if just one step in this process, one structure is missing, the cilium cannot function.

The examples cited above are just three of billions of examples of irreducible complexity in the biological world. Yet, as Behe points out, the proponents of Darwinian evolution offer no explanation for the appearance of irreducible complex structures. The peer-reviewed Journal of **Molecular Evolution has published hundreds of papers on various subjects related to various models of how life could have evolved and possible sequences thereof,** but has published no papers detailing models explaining the intermediate steps by which irreducible complexity arises. And the reason for this, as Behe explains, is simple: Darwinian theory cannot explain it.

Earlier we discussed the incredible complexity of DNA. Human DNA contains about three billion base pairs that have to be in a precise sequence to give rise to a functioning human being. Yet such irreducible complexity does not sway staunch Darwinists like Richard Dawkins, author of such books as *The Selfish Gene* and *The God Delusion*, from their insistence that Darwinian evolution explains the origin and development of life. When confronted with such questions as how did the first living cell appear, Dawkins simply shrugs and says no one knows. On one occasion he conceded that life on Earth may have been seeded here by a civilization much more advanced than ours, but quickly added that somewhere along the line that civilization arose through some Darwinian process. Thus, believers in evolution do not even feel the need to defend the theory. In their view it is established fact, a given.

Consider, then, the dilemma certain scientists have in explaining why the physical constants of nature seem "fine-tuned" to give rise to life. Such scientific luminaries as Stephen Hawking, Lee Smolin, Joseph Silk, Andrei Linde, Leonard Susskind, and Paul Davies, among many others, have noted this troublesome fact. Conveniently, they invoke the "multiverse" theory by way of explanation. The theory asserts that there are many universes, and we just happen to live in one that allows the possibility of carbon-based life. A very convenient theory if one is trying to avoid the possibility that the atom-based matter universe as we know it was designed and engineered by an intelligent designer. But is the idea even logical, or, for that **matter, intellectually honest?**

Many undefined terms get thrown around in physics. For example, consider the term universe. By definition the universe is the totality of all that exists. Physicists are forever coining phrases like "parallel universes" and "hidden dimensions" and, now, "multiverse" to make their theories work. How can we have something parallel to all that exists, or something apart from all that exists, or something hidden within all that exists that is not part of all that exists? The answer to all these questions is: it is impossible.

Ironically, many of the puzzling cosmic coincidences in physics that suggest the possibility of intelligent design were brought to light by scientists who were trying to explain them away rather than adduce them as evidence of intelligent design. The constants of nature seemed to fall within very exact ranges to allow for the emergence of carbon based life.

The gravitational coupling constant would, if slightly stronger, only give rise to stars that were too large and too short-live to support life-sustaining conditions on orbiting planets. If slightly weaker, the gravitational constant would only allow for stars smaller than our sun which would never synthesize elements heavier than helium and spew them into space when exploding as supernovae.

If the strong nuclear force, which holds protons and neutrons together in the atomic nucleus were slightly weaker, hydrogen would be the only element possible. Were it slightly stronger, there would be very little hydrogen in the universe and very little elements heavier than iron.

If the force of electromagnetism were slightly weaker, electrons could not be held in their
orbital shells around nuclei. If the force were stronger, atoms could not share covalent electrons
to form molecules. Further, if the electron to proton mass ratio were different, the electron orbit
would be skewed, making molecules impossible.

Inconveniently, for proponents of the Big Bangig bang theory, the abundance of certain elements like carbon in the universe did not fit the mathematics of their theory. According to Big Bang theory the universe was too young and expanding too rapidly to have produced stars massive enough to have synthesized such heavy elements. This led to the introduction of tweaking factors such as "dark energy" and "dark matter." Yet the average distances between stars seemed just right for the appearance of life. The average distance between stars in the outskirts of our galaxy is about six light years. If stars were closer, planetary orbits would be erratic and hence could not give rise to life. If farther, the heavy elements spewed into space would be too thinly distributed to form planets capable of supporting life.

As we discussed earlier, perhaps the greatest anthropic coincidence was the tri-alpha process by which three helium-4 nuclei are transformed into carbon. The triple-alpha process does not work at the temperatures and pressures assumed to have characterized the Big Bang. Further, no matter what model of the universe one prefers, it is a very exact, complex process requiring a specific sequence, temperatures, and nuclear resonances. It occurs thus: nuclear fusion reaction of two helium-4 nuclei produces beryllium-8, which is very unstable and short-lived – 8.19×10^{-17} seconds – and decays unless in that brief instant a third helium-4 particle fuses with a beryllium-8 nucleus to produce an excited resonance state of carbon-12 which almost always decays back into three helium nuclei, but once every 2421.3 times releases energy and forms the stable form of carbon-12. When a star runs out of hydrogen through fusion into helium it contracts, and this contraction increases its temperature. If the core temperature reaches 10^8 Kelvin then helium nuclei can form fast enough to overcome the beryllium-8 instability factor and become a veritable stable carbon-12 factory.

The triple-alpha process can only occur if carbon-12 and beryllium-8 have nuclear resonances slightly higher than helium-4. In the 1950's this fact was a source of much puzzlement to astrophysicists because based on known nuclear resonances of elements it seemed impossible for stars to produce carbon, or that matter, any element heavier than carbon. In 1953 astrophysicist Fred Hoyle published his calculations that, given the abundance of carbon-12 in the universe, carbon-12 must have a resonance near 7.68 MeV. A subsequent experiment demonstrated that the resonance was only slightly different, 7.65 MeV.

Further studies demonstrated three other uncanny coincidences allowing the synthesis of carbon and other heavier elements. 1) The decay lifetime of a beryllium-8 nucleus is four orders of magnitude larger than the time for two helium-4 nuclei to scatter; 2) If the kinetic energy of the collision of a helium-4 nucleus with a beryllium-8 were any less it could not fuse to produce the excited state of carbon-12, which then can then transition to its stable ground state. The energy level of the excited state of carbon-12 must be between 7.596 MeV and 7.716 MeV in order to produce the abundance of carbon in the universe. Notice that Hoyle's predicted resonance is squarely in this range; 3) In the reaction in which oxygen is synthesized from carbon and helium, the resonance of oxygen in its excited state is such that, if it were only slightly higher, most of the carbon in the universe would have been converted into oxygen.

These cosmic coincidences led Hoyle to argue that the constants of physics had been deliberately fine-tuned by a "superintellect" to bring about intelligent life. Other scientists, perplexed by these coincidences but unwilling to go as far as Hoyle is positing intelligent design,
developed the anthropic principle, which, in its "weak" form simply asserts that the universe appears to be fine-tuned because if did not have certain properties we would not be here. The "strong" anthropic principle has several forms, one of which states categorically that the constants of physics were fine-tuned by, in Hoyle's phrase, a superintellect. Others are less bold, and some rather risible. For example, the "participatory anthropic principle" states that the universe must give rise to intelligent observers because the laws of quantum mechanics demand an observer. The "final anthropic principle" of John Barrow and Frank Tipler argues that the universe is a huge information processing machine and mankind is a part of the effort whereby the universes processes all possible bits of knowledge. In other words, without saying so outright, Barrow and Tipler advance the same

hypothesis as Spinoza: that the universe, in its totality, comprises an intelligent being.

Consider how the strong anthropic principle, particularly the idea of fine-tuning to create just the right conditions for the appearance of carbon-based life, is compatible with zero point theory. If space is the dense plenum of energy described previously, then our theory not only describes what it is that is being fine-tuned, but a possible explanation of the being that is doing the fine-tuning. Further, consider the fact that we as observers have no standard of comparison to judge what other types of cosmological habitats may exist. Our little corner of the universe, a region of space populated by galaxies, stars, and planets composed of elements the nature of which is determined by the number of protons and neutrons in their nuclei and the electrons in the orbital clouds around them, may be a total anomaly, something totally unlike other regions of the universe and unrecognizable to their inhabitants. The assumption that our part of the universe is the norm, and that it appeared out of the fabric of space spontaneously due to the blind
machinations of necessity is an untenable assumption. With all the foregoing examples of design we have adduced, does not the hypothesis that the portion of the universe that we know, at least, and we ourselves, are the product of design answer more questions than it raises? Is it not a more tenable theory than "multiverses" or Darwinism?

The existence of life and complexity itself in the universe is a vexing matter whether or not one believes in intelligent design or not. In order to explain these phenomena thinkers from diverse fields have developed the interdisciplinary fields of emergence, self-organization, and chaos theory. As we shall see, these disciplines do not explain the complexity of living beings or the anthropic coincidences because the physical and chemical processes that give rise to this complexities are complex in themselves. However, ironically, they may explain how the ultimate complexity – intelligence – appeared within the zero point field.

Chaos, Self-Organization, and Emergence

The second law of thermodynamics states that entropy, otherwise known as disorder, will always increase in a closed system, that is, a system in which some form of energy is not continuously introduced. The fact that we have complex systems in the universe demonstrates that the universe is, by definition, not a closed system. If zero point energy is real, and space itself is an infinite plenum of energy, then by definition the second law of thermodynamics does not apply to the universe as a whole. But how does this explain the appearance of complexity in the first place?

Let us consider the hypothesis that the constants of physics were fine tuned to give rise to
intelligent carbon-based life. The raw material which gives rise to the hydrogen which makes up
the newborn galaxies that give rise to the nebulae that give rise to the stars, supernovae, and planets which allow for intelligent life is the zero point field, supposedly an isotropic, homogenous plenum of energy. According to chaos theory, new patterns can emerge within a dynamical system when certain "initial conditions" exist. In his book *The Essence of Chaos* Edward Lorenz explained that certain dynamical systems are "sensitive" to initial conditions, and create emergent properties that would not have appeared save for that initial "alteration."

Emergent properties are simply entities or conditions that differ from, and are unpredictable by, their constituent parts or preconditions. Expressed another way, an emergent property is irreducible to its component parts – as in "irreducible complexity." Some critics of emergence theory complain that it is subjective, that it depends on what a subjective observer considers random or complex. This objection is easily refuted if one assumes that survivability or sustainability of the emergent entity in question is the key criterion for determining its status as emergent. If the configuration survives and, better yet, gives rise to even more complex configurations, it would be reasonable to define it as emergent.

Closely related to chaos and emergence is the phenomenon of self-organization, in which order arises from interactions between parts of an initially disordered dynamical system if sufficient energy is available, self-organization can occur spontaneously. Consider: "parts" of an "initially disordered system" implies a condition that is not homogenous, that is not uniform throughout. Given the principles of chaos, emergence, and self-organization, consider the possibilities of an infinite, dense plenum of energy in which just a small non-uniformity appeared. Infinite time, infinite space, infinite possibilities – infinite complexity.

In our little corner of the universe we see evidence of design. If we accept zero point theory, then the principles of chaos, emergence, and self-organization give a strong clue to the nature and, perhaps, origin of the designer. And perhaps, beyond our powers of imagination, lies something even more fundamental than the zero point field.

But how good it is that we are even here to consider that possibility.

www.ingramcontent.com/pod-product-compliance
Lightning Source LLC
Chambersburg PA
CBHW062343290526
45794CB00005B/2088

* 9 7 9 8 8 6 3 9 7 9 9 1 5 *

DEMOCRATS

2020

AGENDA

A Guide for the American Voter!

by

Shake Spare

1-Read Disclaimer

VOTOPHICS BOOKS

Published by Votophics Media

+234-706-8435-789

Amazon Kindle Edition

DISCLAIMER

This book is completely of the author's creation. It does not represent the party in United States called The Democrats, nor anything or anyone associated with them. It was not approved by any US Political party nor candidate. Also, this was not compiled under anything related to or associated with the Russian Government. And lastly, the author is not a citizen of any country other than Nigeria. Anything supposed to the contrary is FakeNews.

Thank you.

Signed

Shake Spare

DEDICATION

To all American Voters, Lovers and Haters of Trump. Choose wisely in 2020. Keep America Great.

Welfare for illegals paid for by Americans!

Welfare for illegals paid for by Americans!

Welfare for illegals paid for by Americans!

Welfare for illegals paid for by Americans!

Welfare for illegals paid for by Americans!

Welfare for illegals paid for by Americans!

Welfare for illegals paid for by Americans!

Welfare for illegals paid for by Americans!

Welfare for illegals paid for by Americans!

Welfare for illegals paid for by Americans!

Welfare for illegals paid for by Americans!

Welfare for illegals paid for by Americans!

Welfare for illegals paid for by Americans!

Welfare for illegals paid for by Americans!

Welfare for illegals paid for by Americans!

Welfare for illegals paid for by Americans!

Welfare for illegals paid for by Americans!

Welfare for illegals paid for by Americans!

Nothing more...

9-Read Disclaimer

Let China Steal the Jobs!

Let China Steal the Jobs!

Let China Steal the Jobs!

Let China Steal the Jobs!

Let China Steal the Jobs!

Let China Steal the Jobs!

Let China Steal the Jobs!

Let China Steal the Jobs!

Let China Steal the Jobs!

Let China Steal the Jobs!

Let China Steal the Jobs!

Let China Steal the Jobs!

Let China Steal the Jobs!

Let China Steal the Jobs!

Let China Steal the Jobs!

Let China Steal the Jobs!

Let China Steal the Jobs!

Let China Steal the Jobs!

Nothing more...

10-Read Disclaimer

Empty Promises they can't keep!

Empty Promises they can't keep!

Empty Promises they can't keep!

Empty Promises they can't keep!

Empty Promises they can't keep!

Empty Promises they can't keep!

Empty Promises they can't keep!

Empty Promises they can't keep!

Empty Promises they can't keep!

Empty Promises they can't keep!

Empty Promises they can't keep!

Empty Promises they can't keep!

Empty Promises they can't keep!

Empty Promises they can't keep!

Empty Promises they can't keep!

Empty Promises they can't keep!

Empty Promises they can't keep!

Empty Promises they can't keep!

Empty Promises they can't keep!

Empty Promises they can't keep!

11-Read Disclaimer

Empty Promises they can't keep!

Empty Promises they can't keep!

Empty Promises they can't keep!

Empty Promises they can't keep!

Empty Promises they can't keep!

Empty Promises they can't keep!

Empty Promises they can't keep!

Empty Promises they can't keep!

Empty Promises they can't keep!

Empty Promises they can't keep!

Empty Promises they can't keep!

Empty Promises they can't keep!

Empty Promises they can't keep!

Empty Promises they can't keep!

Empty Promises they can't keep!

Empty Promises they can't keep!

Empty Promises they can't keep!

Empty Promises they can't keep!

Nothing more...

12-Read Disclaimer

CHAPTER TWO

HEALTH CARE

Free Abortion, Even to late term.

Free Abortion, Even to late term.

Free Abortion, Even to late term.

Free Abortion, Even to late term.

Free Abortion, Even to late term.

Free Abortion, Even to late term.

Free Abortion, Even to late term.

Free Abortion, Even to late term.

Free Abortion, Even to late term.

Free Abortion, Even to late term.

Free Abortion, Even to late term.

Free Abortion, Even to late term.

Free Abortion, Even to late term.

Free Abortion, Even to late term.

Free Abortion, Even to late term.

Free Abortion, Even to late term.

Free Abortion, Even to late term.

Free Abortion, Even to late term.

Free Abortion, Even to late term.

Free Abortion, Even to late term.

14-Read Disclaimer

Free Abortion, Even to late term.

Free Abortion, Even to late term.

Free Abortion, Even to late term.

Free Abortion, Even to late term.

Free Abortion, Even to late term.

Free Abortion, Even to late term.

Free Abortion, Even to late term.

Free Abortion, Even to late term.

Free Abortion, Even to late term.

Free Abortion, Even to late term.

Free Abortion, Even to late term.

Free Abortion, Even to late term.

Free Abortion, Even to late term.

Free Abortion, Even to late term.

Free Abortion, Even to late term.

Free Abortion, Even to late term.

Free Abortion, Even to late term.

Free Abortion, Even to late term.

Nothing more...

15-Read Disclaimer

Long Waits to see a doctor!

Long Waits to see a doctor!

Long Waits to see a doctor!

Long Waits to see a doctor!

Long Waits to see a doctor!

Long Waits to see a doctor!

Long Waits to see a doctor!

Long Waits to see a doctor!

Long Waits to see a doctor!

Long Waits to see a doctor!

Long Waits to see a doctor!

Long Waits to see a doctor!

Long Waits to see a doctor!

Long Waits to see a doctor!

Long Waits to see a doctor!

Long Waits to see a doctor!

Long Waits to see a doctor!

Long Waits to see a doctor!

Long Waits to see a doctor!

Long Waits to see a doctor!

16-Read Disclaimer

Long Waits to see a doctor!

Long Waits to see a doctor!

Long Waits to see a doctor!

Long Waits to see a doctor!

Long Waits to see a doctor!

Long Waits to see a doctor!

Long Waits to see a doctor!

Long Waits to see a doctor!

Long Waits to see a doctor!

Long Waits to see a doctor!

Long Waits to see a doctor!

Long Waits to see a doctor!

Long Waits to see a doctor!

Long Waits to see a doctor!

Long Waits to see a doctor!

Long Waits to see a doctor!

Long Waits to see a doctor!

Long Waits to see a doctor!

Nothing more...

17-Read Disclaimer

Free Health Care for illegals too!

Free Health Care for illegals too!

Free Health Care for illegals too!

Free Health Care for illegals too!

Free Health Care for illegals too!

Free Health Care for illegals too!

Free Health Care for illegals too!

Free Health Care for illegals too!

Free Health Care for illegals too!

Free Health Care for illegals too!

Free Health Care for illegals too!

Free Health Care for illegals too!

Free Health Care for illegals too!

Free Health Care for illegals too!

Free Health Care for illegals too!

Free Health Care for illegals too!

Free Health Care for illegals too!

Free Health Care for illegals too!

Free Health Care for illegals too!

Free Health Care for illegals too!

18-Read Disclaimer

Free Health Care for illegals too!

Free Health Care for illegals too!

Free Health Care for illegals too!

Free Health Care for illegals too!

Free Health Care for illegals too!

Free Health Care for illegals too!

Free Health Care for illegals too!

Free Health Care for illegals too!

Free Health Care for illegals too!

Free Health Care for illegals too!

Free Health Care for illegals too!

Free Health Care for illegals too!

Free Health Care for illegals too!

Free Health Care for illegals too!

Free Health Care for illegals too!

Free Health Care for illegals too!

Free Health Care for illegals too!

Free Health Care for illegals too!

Nothing more...

19-Read Disclaimer

CHAPTER THREE
IMMIGRATION

Open Borders!

Open Borders!

Open Borders!

Open Borders!

Open Borders!

Open Borders!

Open Borders!

Open Borders!

Open Borders!

Open Borders!

Open Borders!

Open Borders!

Open Borders!

Open Borders!

Open Borders!

Open Borders!

Open Borders!

Open Borders!

Open Borders!

Open Borders!

21-Read Disclaimer

Open Borders!

Open Borders!

Open Borders!

Open Borders!

Open Borders!

Open Borders!

Open Borders!

Open Borders!

Open Borders!

Open Borders!

Open Borders!

Open Borders!

Open Borders!

Open Borders!

Open Borders!

Open Borders!

Open Borders!

Open Borders!

Nothing more...

Legallize illegal immigration!

Legallize illegal immigration!

Legallize illegal immigration!

Legallize illegal immigration!

Legallize illegal immigration!

Legallize illegal immigration!

Legallize illegal immigration!

Legallize illegal immigration!

Legallize illegal immigration!

Legallize illegal immigration!

Legallize illegal immigration!

Legallize illegal immigration!

Legallize illegal immigration!

Legallize illegal immigration!

Legallize illegal immigration!

Legallize illegal immigration!

Legallize illegal immigration!

Legallize illegal immigration!

Legallize illegal immigration!

Legallize illegal immigration!

Legallize illegal immigration!

Legallize illegal immigration!

Legallize illegal immigration!

Legallize illegal immigration!

Legallize illegal immigration!

Legallize illegal immigration!

Legallize illegal immigration!

Legallize illegal immigration!

Legallize illegal immigration!

Legallize illegal immigration!

Legallize illegal immigration!

Legallize illegal immigration!

Legallize illegal immigration!

Legallize illegal immigration!

Legallize illegal immigration!

Legallize illegal immigration!

Nothing more...

No border crisis...(fact checked)...oh there's crisis, it's Trump's fault and we aren't fixing it unless you vote for us...(fact check)

No border crisis...(fact checked)...oh there's crisis, it's Trump's fault and we aren't fixing it unless you vote for us...(fact check)

No border crisis...(fact checked)...oh there's crisis, it's Trump's fault and we aren't fixing it unless you vote for us...(fact check)

No border crisis...(fact checked)...oh there's crisis, it's Trump's fault and we aren't fixing it unless you vote for us...(fact check)

No border crisis...(fact checked)...oh there's crisis, it's Trump's fault and we aren't fixing it unless you vote for us...(fact check)

No border crisis...(fact checked)...oh there's crisis, it's Trump's fault and we aren't fixing it unless you vote for us...(fact check)

No border crisis...(fact checked)...oh there's crisis, it's Trump's fault and we aren't fixing it unless you vote for us...(fact check)

No border crisis...(fact checked)...oh there's crisis, it's Trump's fault and we aren't fixing it unless you vote for us...(fact check)

No border crisis...(fact checked)...oh there's crisis, it's Trump's fault and we aren't fixing it unless you vote for us...(fact check)

No border crisis...(fact checked)...oh there's crisis, it's Trump's fault and we aren't fixing it unless you vote for us...(fact check)

No border crisis...(fact checked)...oh there's crisis, it's Trump's fault and we aren't fixing it unless you vote for us...(fact check)

Nothing more...

MS 13 Gang Members are Human too, not animals!

MS 13 Gang Members are Human too, not animals!

MS 13 Gang Members are Human too, not animals!

MS 13 Gang Members are Human too, not animals!

MS 13 Gang Members are Human too, not animals!

MS 13 Gang Members are Human too, not animals!

MS 13 Gang Members are Human too, not animals!

MS 13 Gang Members are Human too, not animals!

MS 13 Gang Members are Human too, not animals!

MS 13 Gang Members are Human too, not animals!

MS 13 Gang Members are Human too, not animals!

MS 13 Gang Members are Human too, not animals!

MS 13 Gang Members are Human too, not animals!

MS 13 Gang Members are Human too, not animals!

MS 13 Gang Members are Human too, not animals!

MS 13 Gang Members are Human too, not animals!

MS 13 Gang Members are Human too, not animals!

MS 13 Gang Members are Human too, not animals!

Nothing more...

26-Read Disclaimer

CHAPTER FOUR
NATIONAL SECURITY

Abolish ice.

Abolish ice.

Abolish ice.

Abolish ice.

Abolish ice.

Abolish ice.

Abolish ice.

Abolish ice.

Abolish ice.

Abolish ice.

Abolish ice.

Abolish ice.

Abolish ice.

Abolish ice.

Abolish ice.

Abolish ice.

Abolish ice.

Abolish ice.

Abolish ice.

Abolish ice.

28-Read Disclaimer

Abolish ice.

Abolish ice.

Abolish ice.

Abolish ice.

Abolish ice.

Abolish ice.

Abolish ice.

Abolish ice.

Abolish ice.

Abolish ice.

Abolish ice.

Abolish ice.

Abolish ice.

Abolish ice.

Abolish ice.

Abolish ice.

Abolish ice.

Abolish ice.

Nothing more...

Gun confiscation: we will take your guns.

Gun confiscation: we will take your guns.

Gun confiscation: we will take your guns.

Gun confiscation: we will take your guns.

Gun confiscation: we will take your guns.

Gun confiscation: we will take your guns.

Gun confiscation: we will take your guns.

Gun confiscation: we will take your guns.

Gun confiscation: we will take your guns.

Gun confiscation: we will take your guns.

Gun confiscation: we will take your guns.

Gun confiscation: we will take your guns.

Gun confiscation: we will take your guns.

Gun confiscation: we will take your guns.

Gun confiscation: we will take your guns.

Gun confiscation: we will take your guns.

Gun confiscation: we will take your guns.

Gun confiscation: we will take your guns.

Gun confiscation: we will take your guns.

Gun confiscation: we will take your guns.

30-Read Disclaimer

Gun confiscation: we will take your guns.

Gun confiscation: we will take your guns.

Gun confiscation: we will take your guns.

Gun confiscation: we will take your guns.

Gun confiscation: we will take your guns.

Gun confiscation: we will take your guns.

Gun confiscation: we will take your guns.

Gun confiscation: we will take your guns.

Gun confiscation: we will take your guns.

Gun confiscation: we will take your guns.

Gun confiscation: we will take your guns.

Gun confiscation: we will take your guns.

Gun confiscation: we will take your guns.

Gun confiscation: we will take your guns.

Gun confiscation: we will take your guns.

Gun confiscation: we will take your guns.

Gun confiscation: we will take your guns.

Gun confiscation: we will take your guns.

Nothing more...

31-Read Disclaimer

Under fund our military.

Under fund our military.

Under fund our military.

Under fund our military.

Under fund our military.

Under fund our military.

Under fund our military.

Under fund our military.

Under fund our military.

Under fund our military.

Under fund our military.

Under fund our military.

Under fund our military.

Under fund our military.

Under fund our military.

Under fund our military.

Under fund our military.

Under fund our military.

Nothing more...

32-Read Disclaimer

Give Iran more money, so they can shoot down more US drones!

Give Iran more money, so they can shoot down more US drones!

Give Iran more money, so they can shoot down more US drones!

Give Iran more money, so they can shoot down more US drones!

Give Iran more money, so they can shoot down more US drones!

Give Iran more money, so they can shoot down more US drones!

Give Iran more money, so they can shoot down more US drones!

Give Iran more money, so they can shoot down more US drones!

Give Iran more money, so they can shoot down more US drones!

Give Iran more money, so they can shoot down more US drones!

Give Iran more money, so they can shoot down more US drones!

Give Iran more money, so they can shoot down more US drones!

Give Iran more money, so they can shoot down more US drones!

Give Iran more money, so they can shoot down more US drones!

Give Iran more money, so they can shoot down more US drones!

Give Iran more money, so they can shoot down more US drones!

Give Iran more money, so they can shoot down more US drones!

Give Iran more money, so they can shoot down more US drones!

Nothing more...

CHAPTER FIVE

INFRASTRUCTURE

Nothing Here...we aren't just going to help Trump do anything unless...

Nothing Here...we aren't just going to help Trump do anything unless...

Nothing Here...we aren't just going to help Trump do anything unless...

Nothing Here...we aren't just going to help Trump do anything unless...

Nothing Here...we aren't just going to help Trump do anything unless...

Nothing Here...we aren't just going to help Trump do anything unless...

Nothing Here...we aren't just going to help Trump do anything unless...

Nothing Here...we aren't just going to help Trump do anything unless...

Nothing Here...we aren't just going to help Trump do anything unless...

Nothing Here...we aren't just going to help Trump do anything unless...

Nothing Here...we aren't just going to help Trump do anything unless...

Nothing Here...we aren't just going to help Trump do anything unless...

Nothing Here...we aren't just going to help Trump do anything unless...

Nothing Here...we aren't just going to help Trump do anything unless...

Nothing Here...we aren't just going to help Trump do anything unless...

Nothing Here...we aren't just going to help Trump do anything unless...

Nothing Here...we aren't just going to help Trump do anything unless...

Nothing Here...we aren't just going to help Trump do anything unless...

Nothing Here...we aren't just going to help Trump do anything unless...

Nothing Here...we aren't just going to help Trump do anything unless...

35-Read Disclaimer

Nothing Here...we aren't just going to help Trump do anything unless...

Nothing Here...we aren't just going to help Trump do anything unless...

Nothing Here...we aren't just going to help Trump do anything unless...

Nothing Here...we aren't just going to help Trump do anything unless...

Nothing Here...we aren't just going to help Trump do anything unless...

Nothing Here...we aren't just going to help Trump do anything unless...

Nothing Here...we aren't just going to help Trump do anything unless...

Nothing Here...we aren't just going to help Trump do anything unless...

Nothing Here...we aren't just going to help Trump do anything unless...

Nothing Here...we aren't just going to help Trump do anything unless...

Nothing Here...we aren't just going to help Trump do anything unless...

Nothing Here...we aren't just going to help Trump do anything unless...

Nothing Here...we aren't just going to help Trump do anything unless...

Nothing Here...we aren't just going to help Trump do anything unless...

Nothing Here...we aren't just going to help Trump do anything unless...

Nothing Here...we aren't just going to help Trump do anything unless...

Nothing Here...we aren't just going to help Trump do anything unless...

Nothing Here...we aren't just going to help Trump do anything unless...

Nothing at all... except... nothing.

36-Read Disclaimer

CHAPTER SIX
Human Rights

If we don't like it, don't say it! Ban Free Speech!

If we don't like it, don't say it! Ban Free Speech!

If we don't like it, don't say it! Ban Free Speech!

If we don't like it, don't say it! Ban Free Speech!

If we don't like it, don't say it! Ban Free Speech!

If we don't like it, don't say it! Ban Free Speech!

If we don't like it, don't say it! Ban Free Speech!

If we don't like it, don't say it! Ban Free Speech!

If we don't like it, don't say it! Ban Free Speech!

If we don't like it, don't say it! Ban Free Speech!

If we don't like it, don't say it! Ban Free Speech!

If we don't like it, don't say it! Ban Free Speech!

If we don't like it, don't say it! Ban Free Speech!

If we don't like it, don't say it! Ban Free Speech!

If we don't like it, don't say it! Ban Free Speech!

If we don't like it, don't say it! Ban Free Speech!

If we don't like it, don't say it! Ban Free Speech!

If we don't like it, don't say it! Ban Free Speech!

If we don't like it, don't say it! Ban Free Speech!

If we don't like it, don't say it! Ban Free Speech!

38-Read Disclaimer

If we don't like it, don't say it! Ban Free Speech!

If we don't like it, don't say it! Ban Free Speech!

If we don't like it, don't say it! Ban Free Speech!

If we don't like it, don't say it! Ban Free Speech!

If we don't like it, don't say it! Ban Free Speech!

If we don't like it, don't say it! Ban Free Speech!

If we don't like it, don't say it! Ban Free Speech!

If we don't like it, don't say it! Ban Free Speech!

If we don't like it, don't say it! Ban Free Speech!

If we don't like it, don't say it! Ban Free Speech!

If we don't like it, don't say it! Ban Free Speech!

If we don't like it, don't say it! Ban Free Speech!

If we don't like it, don't say it! Ban Free Speech!

If we don't like it, don't say it! Ban Free Speech!

If we don't like it, don't say it! Ban Free Speech!

If we don't like it, don't say it! Ban Free Speech!

If we don't like it, don't say it! Ban Free Speech!

If we don't like it, don't say it! Ban Free Speech!

Nothing more...

39-Read Disclaimer

Equality for women, minus conservative women!

Equality for women, minus conservative women!

Equality for women, minus conservative women!

Equality for women, minus conservative women!

Equality for women, minus conservative women!

Equality for women, minus conservative women!

Equality for women, minus conservative women!

Equality for women, minus conservative women!

Equality for women, minus conservative women!

Equality for women, minus conservative women!

Equality for women, minus conservative women!

Equality for women, minus conservative women!

Equality for women, minus conservative women!

Equality for women, minus conservative women!

Equality for women, minus conservative women!

Equality for women, minus conservative women!

Equality for women, minus conservative women!

Equality for women, minus conservative women!

Nothing more...

40-Read Disclaimer

Babies lives are not human lives!

Babies lives are not human lives!

Babies lives are not human lives!

Babies lives are not human lives!

Babies lives are not human lives!

Babies lives are not human lives!

Babies lives are not human lives!

Babies lives are not human lives!

Babies lives are not human lives!

Babies lives are not human lives!

Babies lives are not human lives!

Babies lives are not human lives!

Babies lives are not human lives!

Babies lives are not human lives!

Babies lives are not human lives!

Babies lives are not human lives!

Babies lives are not human lives!

Babies lives are not human lives!

Babies lives are not human lives!

41-Read Disclaimer

Babies lives are not human lives!

Babies lives are not human lives!

Babies lives are not human lives!

Babies lives are not human lives!

Babies lives are not human lives!

Babies lives are not human lives!

Babies lives are not human lives!

Babies lives are not human lives!

Babies lives are not human lives!

Babies lives are not human lives!

Babies lives are not human lives!

Babies lives are not human lives!

Babies lives are not human lives!

Babies lives are not human lives!

Babies lives are not human lives!

Babies lives are not human lives!

Babies lives are not human lives!

Babies lives are not human lives!

Babies lives are not human lives!

Babies lives are not human lives!

42-Read Disclaimer

Babies lives are not human lives!

Babies lives are not human lives!

Babies lives are not human lives!

Babies lives are not human lives!

Babies lives are not human lives!

Babies lives are not human lives!

Babies lives are not human lives!

Babies lives are not human lives!

Babies lives are not human lives!

Babies lives are not human lives!

Babies lives are not human lives!

Babies lives are not human lives!

Babies lives are not human lives!

Babies lives are not human lives!

Babies lives are not human lives!

Babies lives are not human lives!

Babies lives are not human lives!

Babies lives are not human lives!

Nothing more...

43-Read Disclaimer

Whites should shut up!

Whites should shut up!

Whites should shut up!

Whites should shut up!

Whites should shut up!

Whites should shut up!

Whites should shut up!

Whites should shut up!

Whites should shut up!

Whites should shut up!

Whites should shut up!

Whites should shut up!

Whites should shut up!

Whites should shut up!

Whites should shut up!

Whites should shut up!

Whites should shut up!

Whites should shut up!

Nothing more...

44-Read Disclaimer

CHAPTER SEVEN

FOREIGN POLICY

Just keep apologizing on behalf of America!

Just keep apologizing on behalf of America!

Just keep apologizing on behalf of America!

Just keep apologizing on behalf of America!

Just keep apologizing on behalf of America!

Just keep apologizing on behalf of America!

Just keep apologizing on behalf of America!

Just keep apologizing on behalf of America!

Just keep apologizing on behalf of America!

Just keep apologizing on behalf of America!

Just keep apologizing on behalf of America!

Just keep apologizing on behalf of America!

Just keep apologizing on behalf of America!

Just keep apologizing on behalf of America!

Just keep apologizing on behalf of America!

Just keep apologizing on behalf of America!

Just keep apologizing on behalf of America!

Just keep apologizing on behalf of America!

Nothing more...

46-Read Disclaimer

America last is cool! Let's globalize.

America last is cool! Let's globalize.

America last is cool! Let's globalize.

America last is cool! Let's globalize.

America last is cool! Let's globalize.

America last is cool! Let's globalize.

America last is cool! Let's globalize.

America last is cool! Let's globalize.

America last is cool! Let's globalize.

America last is cool! Let's globalize.

America last is cool! Let's globalize.

America last is cool! Let's globalize.

America last is cool! Let's globalize.

America last is cool! Let's globalize.

America last is cool! Let's globalize.

America last is cool! Let's globalize.

America last is cool! Let's globalize.

Nothing more...

47-Read Disclaimer

CHAPTER EIGHT

RELIGIOUS POLICY

Christians Must Denounce their faith.

Christians Must Denounce their faith.

Christians Must Denounce their faith.

Christians Must Denounce their faith.

Christians Must Denounce their faith.

Christians Must Denounce their faith.

Christians Must Denounce their faith.

Christians Must Denounce their faith.

Christians Must Denounce their faith.

Christians Must Denounce their faith.

Christians Must Denounce their faith.

Christians Must Denounce their faith.

Christians Must Denounce their faith.

Christians Must Denounce their faith.

Christians Must Denounce their faith.

Christians Must Denounce their faith.

Christians Must Denounce their faith.

Christians Must Denounce their faith.

Christians Must Denounce their faith.

Christians Must Denounce their faith.

Christians Must Denounce their faith.

Christians Must Denounce their faith.

Christians Must Denounce their faith.

Christians Must Denounce their faith.

Christians Must Denounce their faith.

Christians Must Denounce their faith.

Christians Must Denounce their faith.

Christians Must Denounce their faith.

Christians Must Denounce their faith.

Christians Must Denounce their faith.

Christians Must Denounce their faith.

Christians Must Denounce their faith.

Christians Must Denounce their faith.

Christians Must Denounce their faith.

Christians Must Denounce their faith.

Christians Must Denounce their faith.

Christians Must Denounce their faith.

Christians Must Denounce their faith.

Nothing more...

God, get outta our government!

God, get outta our government!

God, get outta our government!

God, get outta our government!

God, get outta our government!

God, get outta our government!

God, get outta our government!

God, get outta our government!

God, get outta our government!

God, get outta our government!

God, get outta our government!

God, get outta our government!

God, get outta our government!

God, get outta our government!

God, get outta our government!

God, get outta our government!

God, get outta our government!

God, get outta our government!

God, get outta our government!

God, get outta our government!

51-Read Disclaimer

God, get outta our government!

God, get outta our government!

God, get outta our government!

God, get outta our government!

God, get outta our government!

God, get outta our government!

God, get outta our government!

God, get outta our government!

God, get outta our government!

God, get outta our government!

God, get outta our government!

God, get outta our government!

God, get outta our government!

God, get outta our government!

God, get outta our government!

God, get outta our government!

God, get outta our government!

God, get outta our government!

Nothing more...

52-Read Disclaimer

CHAPTER NINE

EDUCATION

Even if you didn't go to college, you must pay for some lazy ones' college debt!

Even if you didn't go to college, you must pay for some lazy ones' college debt!

Even if you didn't go to college, you must pay for some lazy ones' college debt!

Even if you didn't go to college, you must pay for some lazy ones' college debt!

Even if you didn't go to college, you must pay for some lazy ones' college debt!

Even if you didn't go to college, you must pay for some lazy ones' college debt!

Even if you didn't go to college, you must pay for some lazy ones' college debt!

Even if you didn't go to college, you must pay for some lazy ones' college debt!

Even if you didn't go to college, you must pay for some lazy ones' college debt!

Even if you didn't go to college, you must pay for some lazy ones' college debt!

Even if you didn't go to college, you must pay for some lazy ones' college debt!

Even if you didn't go to college, you must pay for some lazy ones' college debt!

Even if you didn't go to college, you must pay for some lazy ones' college debt!

Even if you didn't go to college, you must pay for some lazy ones' college debt!

Even if you didn't go to college, you must pay for some lazy ones' college debt!

Even if you didn't go to college, you must pay for some lazy ones' college debt!

Even if you didn't go to college, you must pay for some lazy ones' college debt!

Even if you didn't go to college, you must pay for some lazy ones' college debt!

Even if you didn't go to college, you must pay for some lazy ones' college debt!

Even if you didn't go to college, you must pay for some lazy ones' college debt!

Even if you didn't go to college, you must pay for some lazy ones' college debt!

Even if you didn't go to college, you must pay for some lazy ones' college debt!

Even if you didn't go to college, you must pay for some lazy ones' college debt!

Nothing more...

Sexuallize children at public schools.

Sexuallize children at public schools.

Sexuallize children at public schools.

Sexuallize children at public schools.

Sexuallize children at public schools.

Sexuallize children at public schools.

Sexuallize children at public schools.

Sexuallize children at public schools.

Sexuallize children at public schools.

Sexuallize children at public schools.

Sexuallize children at public schools.

Sexuallize children at public schools.

Sexuallize children at public schools.

Sexuallize children at public schools.

Sexuallize children at public schools.

Sexuallize children at public schools.

Sexuallize children at public schools.

Sexuallize children at public schools.

Sexuallize children at public schools.

Sexuallize children at public schools.

Sexuallize children at public schools.

Sexuallize children at public schools.

Sexuallize children at public schools.

Sexuallize children at public schools.

Sexuallize children at public schools.

Sexuallize children at public schools.

Sexuallize children at public schools.

Sexuallize children at public schools.

Sexuallize children at public schools.

Nothing more...

WHERE DO YOU BELONG?

If you really want to vote for anything called Democrat, this is why. If you don't, this is also why!